D0535667

# Writing on Water

Terra Nova Books aim to show how environmental issues have cultural and artistic components, in addition to the scientific and political. Combining essays, reportage, fiction, art, and poetry, Terra Nova Books reveal the complex and paradoxical ways the natural and the human are continually redefining each other.

Other Terra Nova books:

*The New Earth Reader*

*The World and the Wild*

*The Book of Music and Nature*

Terra Nova
New Jersey Institute of Technology
Newark, NJ 07102
973 642 4673
terranova@njit.edu
http://www-ec.njit.edu/~tn/

# Writing on Water

*edited by David Rothenberg and Marta Ulvaeus*

A Terra Nova Book

The MIT Press
Cambridge, Massachusetts
London, England

This book was set in Berkeley Old Style Book by Wellington Graphics in Corel Ventura and was printed and bound in the United States of America.

Library of Congress Cataloging-in-Publication Data

Writing on water / edited by David Rothenberg and Marta Ulvaeus.
p. cm.
ISBN 0-262-18211-4 (hc. : alk. paper)
1. Water—Literary collections. 2. Water. I. Rothenberg, David, 1962– II. Ulvaeus, Marta.

PN 6071.W37 W75 2001
808.8′036—dc21

00-052693

# Contents

© Jerry Uelsmann

*At first was neither Being nor Nonbeing.*
*There was not air nor yet sky beyond.*
*What was its wrapping? Where? In whose protection?*
*Was Water there, unfathomable and deep?*

*There was no death then, nor yet deathlessness;*
*of night or day there was not any sign.*
*The One breathed without breath, by its own impulse.*
*Other than that was nothing else at all.*

*Darkness was there, all wrapped around by darkness,*
*and all was Water indiscriminate. Then*
*that which was hidden by the Void, that One, emerging,*
*stirring, through power of Ardor, came to be.*

*In the beginning Love arose,*
*which was the primal germ cell of the mind.*
*The Seers, searching in their hearts with wisdom,*
*discovered the connection of Being in Nonbeing.*

*A crosswise line cut Being from Nonbeing.*
*What was described above it, what below?*
*Bearers of seed there were and mighty forces,*
*thrust from below and forward moved above.*

*Who really knows? Who can presume to tell it?*
*Whence was it born? Whence issued this creation?*
*Even the Gods came after its emergence.*
*Then who can tell from whence it came to be?*

*That out of which creation has arisen,*
*whether it held it firm or it did not,*
*He who surveys it in the highest heaven,*
*He surely knows—or maybe He does not!*

"The Hymn of the Origins," from the *Rig Veda* X. 129
Translated by Raimundo Panikkar

# Introduction

*David Rothenberg*

In the beginning, all was water. Swirl, wave, swell, crash of ocean upon ocean. Or maybe not. The *Rig Veda* has enough gall to conclude that no one, perhaps not even the creator, can really know or presume to tell.

But no one can doubt the fundament of water, the curiously clear liquid that buoys the whole drift of life, making this planet so interesting to those of us creatures distanced enough to want to look back at it all, to dry off and once again dive in, to bear witness to the results of this endless flow. That cool, quenching taste is the ultimate metaphor of liquid motion, the one feel everyone knows and needs. We're thirsty for change and constancy, for the same rivers through which different waters endlessly flow.

It is silent and booming, placid and rough, random and bound by the laws of chaos and swirl. Silent blanket as snow, solid berg as ice, raging torrent as flood. It is the easiest thing out there in nature to want to describe, because it moves gently in and out of us. We are mostly water. The planet's surface is mostly water. Once a single element among classical elements, today we're supposed to realize that it's always a compound, not a single thing. Still, there's wonder at the fall of rain and the breach in the dam and the cool of the drink.

How fortunate for human life that the stuff exists! All would be dry or unpalatable without it. Well, how could anything be thought of as dry without wetness to compare it to? Inside the sense of water is the very idea of liquidity. Then which came first: flow or stagnancy, identity or motion? Any long gaze at a swirling stream or an imposing glacier can bring the observer swiftly to metaphysics. How can we begin to understand such a world? Simply by attending to it, perceiving it with the greatest attention, drinking it up.

Say anything about water, and you will sense the truth in its opposite. So Heraclitus can be remembered thousands of years later for saying that we may never step in the same river twice. He also said that among those who do step in the same rivers, different waters flow. Was he trying to explain himself, or wanting to become known as a prophet of opposites? Probably neither, because none of his works has survived the inevitable erosion of time—that wasting away from the efforts of pounding water and open air. We've lost the flow of his words. Only fragments remain, and we can be sure that the author of the notion that all is flow and movement was not a composer of aphorisms of detachment.

Water does not divide; it connects. With simplicity it links all aspects of our existence. We feel its many meanings. It is elemental to human perception even though chemistry has long told us how it can be broken down, but that's not to deny the mystery of that eddy and swirl.

There is some science here in our collection—enough to remind you that experiments have been conducted and that we have learned a lot that can be reproduced over years of hypothesizing and testing. Still, science has no general theory of water, nothing able to explain the essential accidents of molecules bumping up against one another, always in different ways. We also include a few plans for what to do with water, an elemental resource and yet another fundamental part of nature threatened due to human ignorance and hubris despite all of the exacting knowledge we have amassed for ourselves. We must do something about it, and we offer just a few clues here.

Most of what we have chosen are stories, poems, and essays that build on some felt insight of a human encounter with water in any of its many forms. The notion of flow is a testament to our recognition that everything always changes, that the world and our selves, our loves, and our friends are ever on the move. The world comes back upon itself like the tides. We live inside cycles, as the rain becomes the fields and the dew goes back to the clouds.

Do we all have clear memories of water? The fresh lift of seawater, the salty taste drying on our skin in the sun. The solid presence of the memories of *thirst,* as a desire that can be *quenched.* What a great word, so singular, so squeezy and tight, from the Olde Englisch *cwencan,* to disappear—as if our thirst ever will, for water, for ideas, for change, for meaning. I still remember the one moment I felt it most: at the end of a two-day hike at age sixteen in the Canyonlands desert, a few hours after our supply of water had run out. It was a long walk in the late morning sun heading home. *There was no water. Not one of us would last.* Following the long march, a drip of paradise waits, just one rusty faucet, the cabin by the trail's end and the asphalt road back to our culture. Never mind the accoutrements; the pipe was calling, valve open, the flow of clear water right over the face. Yes! Cool relief, this world still wants us . . .

It is amazing how many words we have for water and its places. Spring, trickle, cascade, rapid, fall, creek, river, tarn, pond, lake, bay, sound, sea, ocean, rain, sleet, snowfall, snowflake, snowdrift, snowbank, snowmelt, ice-pack, iceberg, glacier, ice cap, mist, fog, vapor, steam, rings of ice guarding the atmosphere as it ends far out in space—just what are those things called?

There is no escaping its metamorphosis from liquid both ways, into solid or gas. It is chemistry that we experience each day. To account for it, in word or image, is the most immediate kind of nature writing there is. What do these responses to the necessary moisture have in common? What kind of soup have we planned here?

The mission at work in *Terra Nova* has never been to separate a natural life from an artificial one, or to argue that a particular sensibility will save our species or our souls. But I have always had faith that there is more out there to see, to drink, to bathe in, and to be immersed in with both joy and peril if we value more highly our perceptions, not just the "facts," about how the world reaches us and why it makes us possible. Choose your rules, your elements, your first principles. But choose them well, so they enable you to inhale a world as real and wonderful and as ambiguous as is the actual truth.

Taste water everywhere, and you will not die of thirst. Feel its omnipresence, and you can always go for a swim. Fill your cup with as many stories as you dare take. Let water work for you.

I suggest something pragmatic here. Art can do something useful, though that has never been its first goal. Think of the river of humanity observing the liquidity around us, and of the continuing immediacy of ancient reports about how the world seems—that sense in which the five classical elements still resonate with our experience. As much knowledge as we have amassed through the ages, we still see in much the same way as did the ancients. If Thales thought the earth to be afloat or suspended in water, it is because many things surprisingly do float in the drink, from human bodies to even ice, which is strangely less dense than what it melts into. It's not true; this planet doesn't drift in water, but the image continues to make sense. Water is so much a part of us that we are inclined to see the world in watery terms. Looking into water as image and metaphor, listening to it as music, feeling it as rushing substance, it quickly seems to contain or be contained in everything. Let it inspire us; let us explain through it. It welcomes us; it makes us possible. It is clear and muddy, nourishing and dangerous.

Let us never forget that it is imagination that makes any progress possible. Someone first imagined water as *the* element, then qualified it as one of several elements, before the notion of element could be fine-tuned over the centuries into something removed from our real experience of the world. Art can guess before it sees and presage what will one day be discovered. As homebound a poet as Emily Dickinson wrote crisply of the icy pull of the Arctic in her poem #525, with an ecological cry for "Lapland's necessity," where "the Hemlock's nature thrives on cold. The Gnash of Northern winds is sweetest nutriment to him." It has been suggested that she could claim having discovered the polar snow line decades before science would admit that nature required such a thing.[1] It took an icebound leap and extrapolation from the snow in the heavy boughs of the evergreens outside her Amherst home to guess how the trees in the far north must feel.

Indeed, as science has become more and more penetrating, it has also receded from our immediate experience of the world. We are taught not to trust what we see or feel or imagine, that these are mere human guesswork, while information, layered upon itself and developed over centuries with meticulous care, can do more and ultimately knows best for us. It might explain why our tongues stick to the frozen flagpole, but never what it feels like. We need more than ever to keep imagining, not to get trapped in what we are taught incessantly about the cycle of things: freeze, thaw, rain, cloud, trickle, flood. Water may not be life, but it lives, moves, carries us along, and churns us to the bank.

We are made of it, we are enmeshed in it, we need it to survive, and it needs us to preserve it. About so much of nature can those same words be said! Still, I don't think the solution to the problem is admonishment, easy moralism on how far humanity has sunk into the muddy depths of ignorance.

We should learn again to love the world, to trust our senses, to swim naked in waters of all temperatures whenever we can. No amount of layers of civilization and information should take such pleasures from us.

## Note

1. See Bernard Mergen, *Snow in America* (Washington, D.C.: Smithsonian, 1997), pp. 21–22.

# I

## *Source and Substance*

John Einarson, *Backcountry waterfall, Hokkaido, Japan.*

# The Rarest Element

*Sidney Perkowitz*

Not long ago a construction crew rolled into my backyard, armed with digging tools, plumbing supplies and four tons of rocks. A week later they had completed a pond, but one with a difference. Lying at the base of a steep slope, it is fed by water rushing down a long artificial streambed shaped into a gentle S curve and lined with every one of those rocks. The water recirculates, pumped back up the slope through a buried pipe. The marvelous result is that from my wooden deck, cantilevered over the site, I experience water in two states at once: one still and placid, disturbed only by the slow-moving fins of goldfish; the second dynamically turbulent, cascading between rocks here, splashing against a boulder there, and producing a complex, paradoxically soothing white noise.

In essence, what I have arranged to observe from my deck are two faces of the peculiar state of matter called liquid—most prominently represented on our planet by water, $H_2O$. Water in nature has drawn artists for millennia, and both its flow and its serenity have borne deep meaning for philosophers. In the sixth century B.C. the Greek philosopher Thales concluded that water is the ultimate substance, the principle, or element, of all things. A century or so later, when the Greek philosopher Empedocles proposed that the complexities of creation required four elements instead of one, the liquid state took its place among them as the element water.

Among all the substances on earth, liquids in general and water in particular are most nearly unique. Although $H_2O$ molecules have been found in clouds of interstellar gas and even within the sun, bulk liquids of any kind are in short supply beyond our planet. With the possible exception of lakes of hydrocarbons—compounds of hydrogen and carbon—that are thought to exist on Saturn's largest moon, Titan, astronomers know of no other cosmic site where liquids fill basins or flow along channels under open sky.

Venus is too hot to support liquid water; Mars is too cold (though the recent NASA Pathfinder mission corroborated earlier evidence that water once flowed there). The planet Mercury is both too hot and too cold: the illuminated half becomes more than hot enough to melt lead, and the dark half gets as cold as minus 180 degrees Celsius (minus 292 degrees Fahrenheit), so water could exist there only as vapor or ice. And only among the moons of Jupiter do astronomers think liquid water may exist at all: some evidence suggests that an ocean is hidden under a layer of ice on Europa, and other signs have recently hinted that liquid water exists on Callisto. As for other liquids, it is thought that both hydrogen and helium lie deep within Jupiter and Saturn, liquefied under the huge pressures exerted by the overlying weight of an entire giant planet. But except for such narrow niches—and contrary to Thales's view—as far as anyone knows, the universe is dry.

Dry, that is, except for the liberal supplies of water on our home planet. That makes the earth virtually the only game in town when it comes to understanding the liquid state of matter. Among earthly liquids, water is the most prevalent, covering more than 70 percent of the planetary surface, and it is by far the most important, for the scientific consensus is that life could not exist without it. Its central role in life arises because water is a prime natural medium for chemical reactions. Its mobile molecules act to diminish the electromagnetic forces that link atoms together, freeing the atoms to combine chemically with other free-floating atoms. According to present thinking, only a watery environment such as the sea could have supported the chain of chemical reactions that formed such elaborate compounds as chlorophyll, DNA, and hemoglobin, and the presence of water is essential for all the ongoing chemical processes of life.

With so much at stake, you would think that modern science would understand water inside and out—and so it does, but only to a point. We physicists think we know every feature of the water molecule: its two atoms of hydrogen and one of oxygen: its boomerang-like shape, with the oxygen at the center and a hydrogen defining each arm; the angle between those arms (104.523 degrees) and the distance between the oxygen and each hydrogen (0.095718 nanometer), reckoned with exquisite accuracy to the nearest quadrillionth ($10^{-15}$) of a meter (though even those numbers change when enough molecules cluster together to form a liquid). We can reel off impressively precise information about the density of water, its melting and boiling points, its electrical conductivity and many other properties. But understanding fails just as the questions get truly interesting. And so, after more than a century of intense study with a growing armamentarium of scientific tools, techniques, and theories, many of the most basic and familiar properties of water remain tantalizingly, and frustratingly, unexplained.

Some of the fascinating behavior of water is broadly characteristic of liquids in general—such as the mysterious fact that water can generate seemingly random swirls of activity as it flows, a kind of turbulent behavior that physicists are barely beginning to grasp. Other inexplicable properties, however, belong to water alone. Unlike other substances, water does not expand as it is warmed from solid to liquid; it contracts. That makes ice less dense than water, a peculiarity whose consequences touch a surprisingly broad slice of human life (it is, for instance, the reason the *Titanic* sank, since it is the reason icebergs float, and it is the reason lakes do not freeze from the bottom up). As water is heated beyond its melting point, it continues to contract until it reaches its maximum density, at four degrees Celsius (thirty-nine degrees Fahrenheit)—a fact that is not incidentally related to the temperatures between two and four degrees Celsius that prevail at the bottom of the oceans, where the densest seawater sinks.

There are other anomalies. It takes a lot of energy to heat water, much more, pound for pound, than it takes to heat many other compounds. The freezing and boiling points of water are high compared with those of similar compounds. It becomes less, not more compressible as it heats up. And, counterintuitively, water under pressure flows not less easily, but more.

All those properties arise from the collective interactions of the water molecules—which would be hard to examine even if water were truly at rest. But, notwithstanding the metaphorical truth of the adage, "Still waters run deep," in fact there is no still water. When I turn off the pump driving my backyard stream, the pond looks absolutely serene under the sun. Far below the limits of human perception, however, its molecules are agitated by thermal energy and thrown together in a mosh pit of motion so complicated its precise dance steps remain a mystery.

One problem is the sheer number of dancers, the swarm of molecules in even the least drop of water. (A mere eighteen grams—about half an ounce—is made up of Avogadro's number of molecules: $6 \times 10^{23}$ of them.) Enormous numbers of atoms or molecules make up solids and gases in bulk, too, of course, yet physicists understand those states of matter much better than they do the liquid state. The difference arises from the strength of the interactions among the molecules. At one extreme are the solids, their atoms frozen in place by potent interatomic bonds that hold them fixed. Aside from some residual wiggling of the solid lattice, that stability eliminates the complications of atomic motion, and physicists and metallurgists can now routinely explain and predict the properties of many solids.

At the other extreme, in a gas under ordinary conditions, the molecules are widely separated, move rapidly and collide violently, though rarely. Each molecule can be treated as an individual, which simplifies things enormously. Furthermore, taken en masse, the molecules of different gases are more or less interchangeable: they all collide and rebound in virtually the same way. That is why physicists and chemists can talk about an "ideal gas" and can describe much (though not all) gaseous behavior by appealing to the simple ideal-gas law.

Liquids occupy the great middle region between those two extremes. Their molecules are neither locked in position nor completely free to roam. Each molecule has plenty of leeway to turn, twist, and vibrate, and as one molecule does so, forces emanating from its electron clouds can brush up against other, neighboring molecules and set them turning, twisting and vibrating as well. Compared with the slam-bang encounters in a gas, however, the interplay of molecules in a liquid is much more intimate, more dependent on the idiosyncrasies of the elements or compounds involved. That is why there can be no "ideal liquid," only a rogues' gallery of misfits in which, unfortunately for physicists, water is one of the tougher customers.

The most tractable liquids, at least when it comes to understanding them scientifically, are the ones made up of symmetrical molecules. Methane, for instance, which becomes a liquid at minus 184 degrees Celsius (minus 299 degrees Fahrenheit), traces out the shape of a regular triangular pyramid with a carbon atom at the center. Best-behaved of all are liquids made of the so-called noble gases such as argon, elements whose electrons are configured in such a way as to resist forming chemical bonds. As a result, argon is made of stand-alone atoms, each surrounded by a tidy, self-contained cloud of electric charge.

In a perfect world those charges would always stay evenly distributed in all directions. Then every argon atom would repel every other argon atom, and argon would never liquefy. In the real world, though, random changes in the electron clouds cause charge to build up momentarily on one side of an atom. That imbalance disturbs the cloud of a neighboring atom, which, in turn, disturbs its neighbors, making each of them slightly positive on one side, slightly negative on the other. Since opposite charges attract, the atoms draw together, and argon can become a liquid. Those shimmering interactions, known as van der Waals or London forces, are so faint that a smidgen of energy is enough to overcome them. That is why argon melts at a chilly minus 189 degrees Celsius (minus 308.2 degrees Fahrenheit) and boils at minus 186 degrees Celsius (minus 302.8 degrees Fahrenheit). The forces are also relatively easy to describe mathematically. Physicists "solved" liquid argon to their satisfaction in the 1970s and moved on to more challenging problems.

If liquid argon is a slow, stately pavane, liquid water is more like a slam dance. The reason is that electrical imbalances are hardwired into the shape of the water molecules. The cloud of electrons in the $H_2O$ molecule tends to be denser near the oxygen atom, making that part of the molecule electrically negative, and so the charge near each hydrogen atom becomes electrically positive. If two $H_2O$ molecules are close together and properly oriented, a positive hydrogen atom in one attracts the negative oxygen atom in the other. Thus a hydrogen atom always intervenes between two oxygens, one in each molecule, to paste the molecules together with a link known as a hydrogen bond.

Although hydrogen bonds are vastly weaker than the chemical bonds that keep the water molecules themselves from flying apart (the $H_2O$ molecule survives temperatures as high as 1,200 degrees Celsius, or nearly 2,200 degrees Fahrenheit), they are powerful compared with the shadowy forces that knit together liquid argon. They are tenacious too. They maintain water ice in the form of a regular crystalline array up to the relatively balmy temperature of zero degrees Celsius. And, remarkably, hydrogen bonds continue to affect the interactions of water molecules even after the molecules turn into steam at 100 degrees Celsius.

The holy grail of modern research into water is to explain its peculiarities. So far, however, there is no universal theory of water, nor does anyone know a master equation for all its properties. Hence, as an alternative strategy, many investigators have turned to computer models to sum up the actions of all the molecules in a sample of water and arrive at their total effect. Dealing with astronomical numbers of molecules, however, taxes even the biggest and fastest computers. For models that demand quantum mechanical calculations, a dozen molecules are enough to push the limits of current machines. Simpler models make room for hundreds of virtual molecules in the computer, but even that number may be too few for a realistic picture of water to emerge.

One reason is the need to account for the so-called edge effects. The forces acting on a molecule in the middle of a bulk sample of water, surrounded by other water molecules, are quite different from the forces acting on a molecule swimming near a boundary, where water gives way to something else. Edge effects—on the water's surface, in a thin film, inside a narrow capillary tube—probably explain the embarrassment of polywater: a supposedly new kind of polymerized water that Russian investigators excitedly reported in the early 1970s, which turned out to be one of the biggest scientific blunders of the past three decades. Similarly, edge effects can drastically distort the properties of the virtual water in computer models.

Another complication of the molecule-by-molecule approach is that water molecules clump together. Physicists think the hydrogen-bonded molecules in liquid water continually link, separate, and link again, on a timescale of picoseconds, forming specific pattems that one investigator has described as "flickering clusters." No one is sure how many molecules typically make up a cluster, largely because it is not easy to make microsnapshots of such temporary molecular structures.

Highly sensitive techniques, for instance, have been developed for examining the atoms in a solid, but those techniques cannot readily be adapted to capture the dynamics of liquid water. X-ray scattering is insensitive to hydrogen and so is useless for studying hydrogen bonding (though it can be quite valuable for investigating the positions of the oxygen atoms). Scattering with neutrons works better than X-ray scattering for studying hydrogen bonding, because a hydrogen atom does affect a passing neutron. But even that method shows only the average positions of the molecules, and so the entire motion sequence is reduced to no more than a blurred summary over time. It is as if one were to study a movie film by stacking its frames and peering through all of them at once.

Lacking better information, investigators have had no recourse but to pick an approximate model and run with it. In a 1996 article in the journal *Physical Review Letters,* G. Wilse Robinson, a molecular physicist at Texas Tech University in Lubbock, and his colleagues described a model that they had built in order to predict the experimental fact that water is densest near four degrees Celsius. Robinson and his colleagues based their model on an idea proposed more than a century ago by the German physicist Wilhelm Conrad Röntgen. Röntgen hypothesized that many molecules in liquid water tend to collect into clusters similar to the clusters common in ordinary ice: triangular pyramids with one molecule in the middle and four others at the vertexes around it. (The existence of such clusters in liquid water has since been confirmed by X-ray and neutron diffraction studies.) In Robinson's model some of the hydrogen bonds between adjacent triangular pyramids bend as the temperature rises. That bending reduces the space between adjacent pyramids by as much as half and even causes some of the pyramids to collapse. The entire liquid thereby becomes a denser material.

In real water, of course, such pyramids are scattered throughout three dimensions, continually forming, collapsing, re-forming, and moving about like chunks of apple in a constantly stirred Waldorf salad. In the original Texas Tech model, however, the water molecules were lined up like beads on a string, and the distances between adjacent pairs of molecules were adjusted so as to be analogous to the three-dimensional case. The advantage of the simplified approach was that it led to equations the investigators could solve exactly, instead of having to rely on approximations or computer models. The numbers that resulted from the simplified approach brought welcome news: the entire string of molecules became most compact when the modeled temperature reached four degrees Celsius—just as the modelers had hoped to show.

The results were controversial, however. At least one group of investigators argued that Robinson's one-dimensional result did not apply to water in the real world. But Robinson and his colleagues made quick work of that criticism by extending their model from one to three dimensions. The new 3-D model, they say, supports their original finding. And the key element to its success in explaining the anomalies of water is that it takes into account the bonds between adjacent pyramid clusters of water molecules.

If water at rest seems full of pitfalls, matters are positively chaotic for water in motion, such as the stream running down my rocky backyard. But the chaos is not universal. In some places the water hurries over the rocks in a smooth and orderly manner, bulging or dipping so as to follow each rock's underlying form. In such so-called laminar flow, water molecules move along parallel paths, forming layers, or laminae, that trace the shape of the rock. The flow is highly determined and varies smoothly in time and space; that is, if the velocity is measured at any given location and instant, the velocity a millimeter distant, or a millisecond later, is not startlingly different.

Elsewhere, however, my stream is turbulent. Observers have noted for centuries that flowing fluids can form seemingly random swirls. Leonardo da Vinci, for instance, was fascinated by the disorderly nature of moving water and often drew its characteristic eddies. Analytical hydrodynamics was born in the eighteenth century, when the Swiss mathematician Leonhard Euler derived equations for the motion of any fluid, provided it is incompressible. Unfortunately, he omitted the effects of friction, with embarrassing practical consequences. When Euler applied his equations to design a fountain for Frederick the Great of Prussia, it failed to work.

Other investigators added frictional effects to give a complete mathematical description of moving fluids. But the formula that resulted, the so-called Navier-Stokes equation, is brutally difficult to solve. And no one so far has been able to answer the fundamental question: How does deterministic laminar flow break into random whorls? I first realized the depth of the enigma years ago at the Los Alamos Scientific Laboratory (now the Los Alamos National Laboratory) in New Mexico, as a student intern in a group studying turbulence in explosions. To my amazement, I learned that this perfectly visible, human-scale effect is as intellectually challenging as, say, quantum mechanics. It was then, and still is, the great unsolved problem of classical physics. It is no exaggeration to say that physicists know more about the structure of subatomic particles than they do about the swirls and eddies of daily experience.

The mysteries of water are deep enough for many lifetimes of scientific exploration, but they hardly exhaust the enigmas of the liquid state. When helium gas is cooled to the inconceivably frigid temperature of four degrees above absolute zero, it too becomes a liquid. But chill it another two degrees, as the Soviet physicist Pyotr L. Kapitsa reported in 1938, and liquid helium becomes a superfluid: it moves without internal friction. Among the bizarre properties of a superfluid is that its affinity for the walls of its container—the effect that, under more familiar conditions, gives rise to the meniscus associated with surface tension—is unimpeded by its internal cohesion and resistance to flow. The result is that a column of superfluid helium, which is as clear as distilled water, will spontaneously flow up the wall. and over the rim of its container.

Superfluidity is a quantum mechanical effect, and so the bulk fluid that exhibits its strange properties is called a quantum liquid. The helium atoms in the liquid merge into a single quantum superparticle, in which the individual atomic interactions that cause internal friction cannot take place. The effect was first predicted in 1924 by the Indian physicist Satyendra Nath Bose and Albert Einstein, from the rules of quantum mechanics that govern assemblies of certain kinds of particles (helium atoms among them).

It is worth noting that the study of so-called Bose-Einstein condensates—though not in states properly called liquid—continues to generate some of the most intriguing physics of the twentieth century. For example, Bose-Einstein condensation makes certain solids into superconductors at low temperatures. And in 1995 Eric A. Cornell of the National Institute of Standards and Technology in Boulder, Colorado, and Carl E. Wieman of the University of Colorado, also in Boulder, cooled 2,000 metallic atoms to within several billionths of a degree of absolute zero, uniting them into yet another kind of Bose-Einstein condensate that many physicists regard as a new state of matter.

Thoughts of such intellectual ferment, swirling around the unsolved problems of turbulence, quantum fluids, and the perhaps still-undiscovered varieties of bulk matter, fill my mind with wonder as I watch the new stream in my backyard and listen to the murmur of its self-replenishing waters. And that serendipitous intersection of the concrete world with the world of ideas makes me smile. If something as commonplace as a trickle of water can harbor such a mystery, surely the river of undiscovered knowledge will not soon run dry. I am also glad that Thales was wrong—that water is not the secret of the universe, but only a single item in a vast inventory of secrets. Perhaps further exploration will confirm that water as it exists on earth, cool and supple and soothing, makes up only an infinitesimal fraction of a universe that is mostly arid. In that case, so much the more reason to be thankful for the liquid mysteries that make our home planet a first-class cosmic attraction.

*Tidal edge of restored Crissy Field salt marsh.* © Melissa Nelson

# Constructing a Confluence

*Melissa Nelson*

In late August the temperature reached 104 degrees, and I had been running over hot river rocks all day. It was approaching late afternoon, and the sun was less intense, angling in from the southwest. I sat down next to the river across from one of my favorite redwood trees growing out of the rocky bank. Depending on how much rain we had received the previous winter, sometimes this redwood's roots would be exposed to the summer sun; other times, like that year, they would be completely immersed in the cool emerald water of the south fork of the Eel River in northern Mendocino County.

I walked into the river slowly, the cool water climbing up my slightly sunburned skin. Finally submerged, I swam up the shallow current and held onto mossy rocks, letting the river rush over me. After flowing in the invigorating current a while, I let go and floated down into the deep, shadowed pool next to the redwood. This fog-loving tree stood strong and still in the summer heat, exuding a sweet, spicy aroma. I wondered about the amount of water it must need to absorb to moisten and cool the upper part of the tree, and what miraculous inner magic it must perform to reverse gravity and move water at least two hundred feet up into the sky. To appease my curiosity about this vital roots-river contact, I dove into the deep, dark pool to discover what I could of this magnificent tree's submerged roots. Going down only about four feet, I could barely make out the ghostly arms of the redwood roots covered in moss and webs of old twigs and leaves. The water was murky, and small fish darted in and out of sight. I grabbed hold of a couple of slimy gnarled roots, closed my eyes, and began to sway with the deep currents in a type of spontaneous underwater root dance. Soon my lungs were burning for air, and I let go of the roots so I could slowly float to the surface. But as my buoyancy pulled me upward, something else was pulling me down. My heart raced as I realized that my long brown hair had become entwined with the roots. I panicked as I yanked at my hair, scratching myself on the roots. My chest was screaming for air, and I finally ripped off a big clump of my hair and surged to the surface with a large gasp.

Fortunately, this experience did not make me fear the river or the redwood. Instead I became more aware of the dark mystery and strength of underwater life. My love for the river grew in a new way, enhancing my appreciation of its wild beauty and its relations with so many other life forms and processes. Even now, twenty years later, when I see that tree, I have the sense that some of my hair might still be entangled in those redwood roots, perhaps feeding the tree in some way, as the river has fed me with fresh water, salmon, sensuality, and solitude.

I have continued to spend my summers on the Eel River, experiencing the power of its unpredictability. Where will the good swimming hole be next year? Will the gravel beds be high on the west or east bank, or will they spread thin down the entire riverbed? Will the otters still play under the bridge by the serpentine outcropping? Not being a hydrologist, predicting these changes is simply an exercise of imagination.

After living next to diverted, channeled, concretized, contaminated, and predictable water systems in other parts of developed California, I realize how important such wild unpredictability is for ecosystem health—the necessity of a confluence of many freshwater sources, rainfalls, native plants, fish, and wildlife, unmined gravel beds, along with local cultural values and community support to ensure that a river remains a healthy, thriving watershed. I have become committed to helping restore that wild health in the watershed closest to me—an undiverted urban freshwater system of springs and a stream that will be flowing into a restored salt marsh at Crissy Field on the eastern side of the new 1,480-acre Presidio National Park in San Francisco, a recently converted military base.

### *From Springs to Sea*

For the past four years I have been involved in the research and planning for the ecological restoration of the second main urban watershed resource in the Presidio National Park, Crissy Field, and its upland drainages, Tennessee Hollow and El Polin Springs. This watershed consists of five critical interrelated habitats that are disappearing statewide: natural springs, riparian corridor, freshwater wetland, brackish marsh, and tidal salt marsh. Usually a watershed would be restored from its freshwater source to the bay or sea, but for political reasons, the National Park Service (NPS) and other agencies have prioritized this project in the reverse direction, from tidal marsh up to freshwater sources.

El Polin Springs and two other drainage systems create Tennessee Hollow, which flows north under the Presidio and eventually into east Crissy Field. This area is being restored into a 145-acre parkland playground by the Golden Gate National Recreational Area (GGNRA) and its nonprofit sister organization, the Golden Gate National Parks Association (GGNPA). According to these two conservation organizations (one federal, the other private), within three years, with $25 million and a gaggle of hydrologists, civil engineers, ecologists, planners, and conservation politicians, Crissy Field will be transformed from a toxic military waste site filled with debris and cement into a 145-acre park area, including a 20-acre restored salt marsh.

My role in this restoration planning process has been to educate myself, the NPS, and associated agencies about the prehistoric archaeological sites that exist in Crissy Field. This area is considered sacred to many of the local Ohlone people, and the consultation requirements of federal laws, such as the National Historical Preservation Act, need to be met. I have facilitated opportunities for local Ohlone community members to speak directly to park managers to voice their concerns and perspectives on restoring Crissy Field and protecting their cultural resources.

How do aquatic habitats and other factors—local cultural values, archaeological resources, endangered plant species, and landfills—affect water quality, habitat restoration, and community participation? For example, could local California Indian community members co-manage this watershed and eventually harvest native plants for traditional uses such as basket weaving? Could special native plants with bioremediation qualities be used to help clean up the toxic substances from the landfill?

This is the story of the triangular relationship of the natural landscape, the indigenous cultural landscape (pre-1770 and continuing today), and the historical cultural landscape (post-1770 and continuing today) at the San Francisco Presidio. How these three areas and their various components interact and come together is determining the type of restoration and education that is taking place in this new urban national park. In this research and planning process, critical information is being gained that can be used for other restoration projects and for general public education.

I awake to the loud "thwap, thwap" of our African-tapestry curtain pounding against the open wet window. Outside I see Monterey pine trees dancing wildly in this early February storm. "Oh no," I think. "The Native American Storywalk will probably be canceled for today." I get out of bed anyway, knowing that Otis, a Kashaya Pomo elder and cultural educator, will be at the NPS Visitors' Center, rain or shine. I hope a few other folks will weather the storm to experience one of the rare NPS educational programs in which park rangers and interpreters partner up with a Native American leader to talk about Native history in the park.

I rush into the Visitors' Center fifteen minutes late and see that the storm has definitely deterred an audience, but Otis is at the chalkboard with Lisa Hillstrom, a park ranger, translating her environmental song into the Kashaya Pomo language. I learn that river is *bi da.* The verb follows. I learn that for the Kashaya Pomo, a river is *bi,* or sand, and *da,* which means flowing or moving. "Sand flowing" implies river, so the concept of river or flowing water emerges out of the earth context of sand flowing. Soon a couple of families arrive, and we all spend the morning together listening to Otis tell traditional California Indian stories about how Lizard made the sun stay up in the sky and when Elk and Rabbit traded ears for antlers.

Returning home later, I looked up the English word *river* and discovered that it is derived from the French *riviera,* which refers to a bank or edge. The French is thought to come from the Latin, *rive,* for bank or shore. So the English also refers to a river by describing its earth context, in this case the edges or banks of the river, not the river bed, the "sand moving." None of these words refers to water directly, perhaps because its fluid nature resists definition. Maybe a word like *watershed* has become so popular because it more clearly describes a wide variety of land areas dominated by flowing surface water, on which all creatures vitally depend and which we value as a source of sustenance and beauty. The mysterious origins and mythical stories surrounding water and the names it has been given have been with us for a long time. Perhaps because water is so mutable and slippery, it is easiest to talk about in terms of its many cultural meanings and manifestations.

### *Mythical Waters*

I work and live near one of the last freshwater springs in the city of San Francisco. Its name, El Polin Springs, comes from a Spanish word for pollen or seed, representing fertility. A legend mentioned in Presidio history books, and which also appears on an educational plaque near the spring, states that women who live by it, as I do, or drink its water will be blessed with fertility and "eternal bliss." Although park historians have referred to this as an urban legend, and therefore quite recent in origin, they have also stated that the El Polin Springs area contains some of the park's rarest and most sensitive archaeological sites, indicating a strong cultural history. But what were the human uses of and beliefs about this spring? What did the original inhabitants of this area, the Ohlone, believe about this spring, and what do the contemporary Ohlone descendants think today?

Water, a symbol of the original primal waters that gave birth to life on earth, has long been closely associated with women and fertility, through their monthly menstrual flowing. The jars, baskets, and other containers that women have historically used to gather and carry water symbolize the womb as the sacred place of conception. Water has also been associated with the "feminine" principles of receptivity, vulnerability, sensuality, motherhood, nurturing, flowing. In ancient China, "womankind symbolized the great water cycle that lifted the moisture from seas and lakes, transmuted it into clouds and mists, and spread it fruitfully into the dry soil."[1] In Chinese and East Indian, as well as many other tribal folklores, water is also linked to sexuality, not just the fluid sexuality between humans, but the sexuality that Terry Tempest Williams refers to in her writings on "erotics of place" or what Gary Snyder has referred to as "pan-sexuality"—the energetic and physical comingling with the natural landscape. Perhaps because our very bodies, like the earth, are mostly water, we can easily lose sense of our boundaries when we're in contact with water and feel our bodies in flowing sensuality with rivers, lakes, creeks, oceans, bays, rivulets, and rain. Deeply sensuous, water covers whatever it touches. It also cleanses and heals. In the words of water photographer and activist Jennifer Greane, "water selflessly surrounds whatever is dropped into it." Or as the legendary Taoist Lao-tzu wrote in the *Tao Te Ching*, "when the land was ruled with wisdom," the leaders were "selfless as melting ice."

The River Ganges in India, the end point for many sacred river goddesses, is considered the Mother Goddess of purification. Rivers are "liquid Shakti," moving creative forces that protect, cleanse, and nurture. Hindu texts contain many stories of rivers. For example, the Narmada is said to have been created through the sweat of the god Shiva—perspiration from his ascetic practices, or the sweat of his passionate lovemaking with his consort, Paravati. Because of this sacred origin, all of the rocks in the Narmada are considered holy Shiva linghams. Like the Ganges, the Narmada cleanses and purifies everything she touches, even though today she herself is polluted by industrial runoff, pathogenic bacteria, and garbage. How can this be? How can something so sacred be polluted? Religious scholar Donald Swearer has commented that the Hindu belief that the "Ganges will always purify" has been invoked to support anti-environmental practices, with the thought that "Mother will always clean up after us" and therefore there is no need for us to clean up after ourselves. How are we to help heal and restore health to a natural system whose accompanying cultural belief system believes it to be pure? Fran Peavey, director of the Crabgrass Foundation, Women and Water, and other nonprofit organizations, has been working for eighteen years to clean up the Ganges. She does this "by listening to the river and the local people."

### *The Natural Laws of Water*

The environmental predicament we face of diminishing freshwater—in both quantity and quality—is a severe global problem. Of all of the water on earth (70 percent), only 2.5 percent is freshwater. Of that 2.5 percent, two-thirds is locked up in ice caps and glaciers, which means that less than 1 percent of the world's water is freshwater, covering only about 1 percent of the earth's total surface. Although freshwater is renewable, thanks to the solar-powered hydrological cycle, it remains a scarce and precious natural resource. According to the latest reports from the World Resources Institute, freshwater ecosystems have lost a greater proportion of their species and habitat than ecosystems on land or in the oceans. There is no substitute for freshwater except desalinized seawater, which is extremely costly to produce. As Oren Lyons, Onondaga Faithkeeper for the Iroquois Confederacy says, if you want to know what natural law is, look to water. We humans cannot change our thirst and need for water; it is out of our jurisdiction. Our need for water is a natural law that is governed by other forces.

In the United States about 45 percent of our freshwater goes to industrial use, with slightly less, 42 percent, going to agriculture, and the balance to domestic use. Within California, however, 80 percent of the water goes to agriculture, 16 percent to urban use, and the remaining 4 percent to recreation, wildlife, and power generation.[2]

There are approximately sixteen rivers that flow into the Sacramento–San Joaquin River delta, which then flow into the San Francisco Bay and eventually into the Pacific Ocean. This bay-delta system is the largest and perhaps most important estuary on the west coast of North America. Covering such a large surface area (it drains 40 percent of California), the delta receives runoff from industrial, chemical, agricultural, mining, and urban activities and transfers it into the bay and then into the Pacific Ocean. Of all of this water, which would naturally flow into the bay, 50 percent is diverted to other parts of the state. Since the "discovery" of California 150 years ago, one-third of the bay has been diked or filled, and the estuary has lost 95 percent of its tidal wetlands.[3] What were once tidal wetlands are now ports, garbage dumps, airports, industrial parks, malls, farms, and military bases.

Urban development has resulted in a devastating loss of riparian and salt marsh communities in the San Francisco Bay Area. At the Presidio, the nearly two and a half centuries of military activity have severely damaged and polluted the natural communities, especially the water resources. Its conversion from a military base to a park and the Bay Area's long history of public support for protecting and restoring open spaces and native habitats offers a rare opportunity to restore water ecosystems, from bay or ocean and salt marsh up to freshwater source.

### *Human History in the Watershed*

Oral history, ethnographic records, archaeological studies, and predictive models indicate that for thousands of years, the Presidio was inhabited and actively used by its original inhabitants, the Ohlone. The Presidio and the El Polin Springs area within it were "discovered" and colonized by Europeans in 1770, first by Spain, then by Mexico, and, after the discovery of gold in California in 1849, by American colonists. The freshwater springs were highly valued and used by all members of these four distinct nations.

There are at least seven documented prehistorical archaeological sites in the Presidio. Two shell deposits have been located in the El Polin Springs/Tennessee Hollow drainage. And although more investigation needs to be done to determine the exact extent and age of these shell deposits, cultural resource monitor Ayapish Slow of the Chumash Elders Council walked the upper watershed area with me and surmised from the level of decay of the shell deposits in the soil that these were prehistoric deposits.

In addition to restoring the native ecology of this watershed, I am working with local Ohlone descendants and other California Indians to restore the cultural history of this site. The NPS has acknowledged that the Ohlone were actively using and living along the bay edge of Crissy Field: "The Ohlone Indians who inhabited the Bay Area for more than 1,500 years hunted and gathered on land that is now part of the Presidio. Ancient Ohlone middens have been found along the bay near Crissy Field."[4] If the Ohlone were actively living in and using the Crissy Field area, common sense tells us they were also using the freshwater sources, such as El Polin Springs.

The natural environment of this aquatic habitat has been altered by many human hands over the years, from the Ohlone to the U.S. Army most recently. The ecosystem to which this watershed will be restored is itself culturally constructed. Most restoration ecologists aim to restore degraded landscapes "back" to precontact native conditions. This is often called a reference ecosystem because it refers to a time before the land was invaded by foreign plant and animal species. It is thought to be "pristine" and "natural" because, according to this line of thinking, humans did not alter or profane the wilderness. But the latest research in environmental history and indigenous horticulture is revealing (which Native Americans have said all along) that the so-called precontact reference ecosystem is indistinguishable from that which bore the effects of the Ohlone living on the land. It is virtually impossible to distinguish what ecologists perceive as a "natural" precontact reference ecosystem from an indigenously managed cultural landscape. Consequently, restoring the cultural uses and history of this site is a way literally to "re-story" the land with important information about sustainable human-environment relations.

*Crissy Field Channel opening ceremony, November 1999, Presidio National Park, San Francisco.*

### *Behind the Springs: Endangered Serpentine Grassland*

Behind the northeast-facing bluff where El Polin Springs emerges is a large serpentine outcropping known as the Quarry or, more popularly, Inspiration Point. Serpentine is California's state rock. It creates a rare thin, ultra-basic soil type that is low in calcium, high in manganese, and with very high levels of nickel and chromium, which are often toxic to living organisms. This uncommon soil type gives rise to unusual plant communities that are adapted to its unique chemistry. Some of the plants that grow on serpentine are called serpentine endemics because they are found nowhere else. Common plants that prefer more normal soils can sometimes survive the unique conditions of serpentine, but their size is stunted. Serpentine has a greenish-blue color with streaks of black and gray, visible from roads and trails within the Presidio and all around California and the Pacific rim where there are fault lines. Inspiration Point is one of the only serpentine grassland habitats in the GGNRA, and it is the only habitat for the Presidio clarkia *(Clarkia franciscana)*, an endangered plant species. Under the Endangered Species Act, this site is being restored and monitored by both the U.S. Fish and Wildlife Service and the Natural Resources Division of the NPS.

The Ecology Trail, a park trail that earlier cut through the bottom of this outcropping and roughly divided the rare serpentine grassland habitat from the El Polin Springs source, was moved farther north in spring 1998 to avoid affecting the endangered grassland. There is a sloping hill covered with Monterey pine trees at the northeast end of the new Ecology Trail, approximately one hundred feet from the serpentine grassland. The pine trees are between fifty and one hundred years old, and since Monterey pine did not naturally occur in the Presidio, this forest is considered to be nonnative by NPS's Natural Resource Division. Because they shade the clarkia, they have a negative impact on the native grassland.

Many of the pine trees are protected as part of the cultural landscape because they were planted in 1883 as part of the Presidio Forest Plan by Major Jones of the U.S. Army. This historical tree planting effort is one of the reasons that the Presidio was made a National Historic Landmark. Fortunately, however, the Endangered Species Act requires that the habitat of federally listed species such as the clarkia and dwarf flax be protected and enhanced. To balance the dual purposes of natural and cultural preservation, the Natural Resources Division has removed the nonnative pine trees that directly shade and impede the serpentine endemic's recovery.

The protected, culturally significant pine forest area to the north of the grassland was one of several possible sites for a proposed Indian village, originally supported by the California Indian Museum and Cultural Center (CIMCC) in 1997. The village plans include a permanent, partially subterranean traditional roundhouse, a few smaller tule and redwood huts, and a shade arbor. This village would be used for educational purposes and serve as a central meeting point for local Indian people to gather for events and meetings and to conduct food-collecting and basket-weaving demonstrations. This pine forest was the preferred site for the Indian village, but the Natural Resources Division was concerned that the village project could attract too many people to the area, especially children in educational programs, who might inadvertently harm the sensitive serpentine grassland. For this and other reasons, the Indian village project proposal has been dropped from NPS project review status. The main reasons were that the California Indian Museum and Cultural Center made finding a Presidio building for their actual museum their priority, and the Presidio went through a transition from NPS management to the Presidio Trust, a new federal executive agency mandated by law to achieve financial self-sufficiency for the park by 2013. Since it became the lead agency managing the Presidio in July 1998, the Presidio Trust has made the leasing of existing buildings to generate revenue its priority, not the creation of any new structures such as an Indian village.

### *Historic Garbage and Toxic Runoff*

The other water source for the El Polin Springs's freshwater is a seep to the east of the main spring. This secondary source is directly below a historic landfill, Fillsite 1, which contains lead, cyanide, and other toxic materials. Due to the Base Realignment and Closure Act (BRAC) of 1988, the secretary of the defense recommended closing the Presidio as an army post and transferring it to the NPS. This is the first time in the United States that a military base has become a national park. BRAC mandated that the army be held responsible for the environmental cleanup and restoration of the site, so the army set up an

Environmental Restoration Program and Restoration Advisory Board to oversee the cleanup of toxic areas in the Presidio. Fillsite 1 and Landfill 2 are priority areas because they contain so many toxic areas located under a playground, baseball field, and recreational picnic area. When the army revealed its initial "cleanup plan" in the fall of 1997, the Presidio and neighboring communities were shocked to learn that the primary plan of action was no action: the army would simply leave toxic waste buried in the Presidio and monitor toxic levels without removing or detoxifying identified sites.

The NPS and local organizations and environmental groups condemned the proposal and demanded higher cleanup standards for a new national park. Many local community members, myself included, wrote letters of protest about the proposed cleanup process. The people made their point loud and clear: "No action is not an option for a national park!" I and others suggested using bioremediation and phytoremediation technologies used by EPA research scientist Rufus Chaney and biologist and environmental engineer John Todd. "Using nature to heal nature" by researching and selecting specific plants that actively take up toxic materials and break them down into benign materials, Todd and Chaney, with wastewater and soil treatment, respectively, have successfully completed many projects around the country.[5] Cleaning up toxic wastes in this benign biological manner seems to be an appropriate method to adopt for the Presidio national park, which is supposed to be an urban laboratory committed to sustainable solutions to environmental, economic, and social problems. There has been resistance, earlier from the army and now from the Presidio Trust, to use some of these methods.

In May 1999, the toxic cleanup responsibility for the Presidio, along with $100 million, was transferred from the U.S. Army to the Presidio Trust. The trust has promised to clean up the Presidio at least two decades sooner than the army had planned. If this happens, it will greatly expedite the restoration process and ensure clean water and soil for El Polin Springs and the whole watershed.

### *A Riparian Corridor Through Military Artifacts*

El Polin Springs flows down a corridor occupied by former military housing, other buildings, and partially built and natural landscapes. Parts of this corridor are relatively "natural," meaning that surface water runs through an open green space planted with Monterey cypress where sharp-shinned hawks live. The area looks green, but the water does not touch the soil; it is contained within metal drainage pipes and cement ditches. At many points, it runs underneath roads and parking lots. But it does emerge, just next to my office at the Cultural Conservancy, into a small riparian area with six planted redwood trees and some exotic acacias covered with German ivy. Then the spring water disappears again under the fire station parking lot and Doyle Drive, the seismically unsound freeway that connects the Golden Gate with downtown San Francisco. This area is particularly tricky because it has several buried fuel tanks that contain known and unknown waste products. The water's short but intense trip under, over, and through military artifacts finally ends at the east end of Crissy Field. If and when the cleaned-up freshwater flows out of El Polin Springs, the plan is to replace these military artifacts and transform the main stretch of the watershed that connects the springs to the marsh by planting live oak, California bay, and riparian plants such as willow and sedge.

### *Constructing a Confluence: The Crissy Field Marsh*

Coastal wetlands are driven by the ebb and flow of oceanic tides and thus lunar cycles. They are bounded on the land side by extreme high-water monthly spring tides and on the ocean side by the outer edge of the continental shelf, or, in this case, by the San Francisco Bay. Tidal marshes are a type of coastal wetland that includes salt marshes, brackish marshes, and tidal freshwater marshes. Such is the complex interrelated water system being restored at Crissy Field. Tidal marshes need to be protected from the constant wave action of the sea and consequently are almost always associated with a beach barrier—a narrow strip of dunes or beach that protects the marsh from the open sea or, in this instance, the bay. The existing acres of broken asphalt, rubble, and degraded sand dunes will be replaced by a restored, replanted, and expanded area of sand dunes and twenty-acre tidal marsh. Surrounding the marsh and dunes will be a thirty-acre promenade and beach area, a twenty-eight-acre grassy field, and twenty-two acres of picnic areas, visitor amenities, and parking. In the fall of 1999 the tidal marsh was ceremoniously opened, and as of the fall of 2000, most of the native planting has been completed. By spring

2001 the restoration work will be completed, and the new Crissy Field environmental education center will be opened to the public. The GGNRA predicts that 6.5 million visitors will come to the Presidio each year.

The sand dunes flanking the beach at Crissy Field were once part of a vast dune edged by lush salt marshes and lagoons. Historical maps and photographs show the complex contours of sand, water, marsh, tule, dune swales, and tides that defined this rich wetland system. Once it is restored, this area will support the only native foredune community in San Francisco. The Crissy Field tidal marsh will be flushed constantly by the daily rise and fall of tides in the bay and the influx of freshwater from the Tennessee Hollow drainage. This mixing zone between fresh-and saltwater is a complex ecological matrix with dynamic fluctuations in salinity, temperature, nutrients, turbidity, and sediment load, and it is the habitat for unique aquatic biodiversity that was part of the subsistence economy of the Ohlone peoples.

Before contact with Spanish colonists, the Ohlone fished in the lagoons and on the bay, gathered mussels, clams, and other seafood, and used the thick tule, cattail, willow, and other plants for food, medicine, and household items such as baskets and sleeping mats. Tule boats were built and used to travel around the bay and out to the middle of the marsh, where there was a stabilized sand dune, today called Sand Island or Strawberry Island. The fact that at least two burial sites were found there indicates that the area also had special religious significance.

When the Spanish came, all of this changed dramatically. Wetlands were seen as "stinking swamps" that should be drained, filled, or used for garbage and sewage. Historical research on the Crissy Field marsh indicates that it was a relatively shallow marsh in the late 1800s when the U.S. military began filling in this area. In the early part of the 1900s, huge amounts of sand and fill were brought in to make the wetlands solid ground for the Panama-Pacific World's Fair. Sadly, this diverse aquatic ecosystem and bird habitat was destroyed.

While preparing for the World's Fair in 1912, workers discovered a large shell mound at Crissy Field. An antiquarian named L. L. Loud did a partial excavation in the shell mound area that had not been submerged by the bay waters coming in as a result of the army's filling activities. Interestingly, the Archaeological Site Survey Record mentions "burials recovered," but Loud's actual site description does not repeat that information. An article that appeared in the *San Francisco Examiner* in 1912, however, corroborates the survey record, stating that "human bones in a fair state of preservation were un-earthed."[6] Unfortunately, no further records can be found tracing what happened to these human remains. Today's NPS archaeologists have selectively concluded that because Loud did not mention it in his description, there was not a burial in the shell mound. But even a cursory examination of the shell mounds around the Bay Area suggests a high correlation between shell mounds and human burials. This supposition was supported in 1972 when human remains were discovered in Crissy Field in an area within 150 feet from where the shell mound was discovered.[7] This site contained a human skeleton and a cut mammal bone tube made from a bird radius, most likely used as a whistle. The San Francisco State University Anthropology Identification Lab identified the burial to have been that of a twenty- to thirty-year-old California Indian woman who died in approximately 740 A.D., nearly 1,300 years ago.[8] It would be sound judgment, then, to postulate that Crissy Field had special significance to the Ohlone inhabitants who buried their dead there with cultural items used for ceremonies and healing purposes.

As recently as July 20, 1998, another prehistoric archaeological site was identified in the area of Crissy Field where the historic salt marsh was located. As part of an NPS archaeological investigation, the Army Corps of Engineers and their contractors had been preparing to do more ground excavations in this area to remove contaminated soils and prepare the site for the restored tidal marsh. In the excavation process they discovered "a very dark grey/black shell midden with clams, mussel, cockle, other shellfish; fish, bird, mammal, sea mammal bones; fire-effected rock; Franciscan chert flakes and tools; two obsidian artifacts recovered, including a large serrate point."[9] The initial carbon dating of these materials revealed an age of well over 1,200 years. All of this evidence indicates that what we have been calling Crissy Field for a mere eighty years was for thousands of years a carefully managed cultural landscape of the Ohlone peoples.

With all of this rich history in mind, it is important to realize that what the NPS is deciding to restore at Crissy Field (and other parts of the Presidio) is truly a selective historical landscape. Their restoration construct is based on two ideologies, one of which holds that the wetlands was a "natural" wilderness area "untouched" or at least unmodified by human hands. The other values and prioritizes an eighty-year-old military history (Crissy Field as a World

War I airfield) above thousand-year-old examples of human societies living sustainably in biologically diverse landscapes. This is not just an arbitrary preference for a park tourist destination. There are historical, ethical, and legal processes that have contributed to the NPS's and now the Presidio Trust's decisions. But when we erase human hands from the natural landscape, we are striving for an unrealistic ideal of purity and compromising our humanity.

Fortunately, through years of education and negotiation between community organizations like the Cultural Conservancy, local Ohlone leaders, other Native Americans, and the NPS, a historic memorandum of understanding was signed in July 1999 between the NPS and five local Ohlone leaders, designed to protect Native American cultural resources at Crissy Field, comply with necessary laws and policies regarding agency consultation with Native Americans, and collaboratively develop public education and interpretation programs that honor the history and contemporary lives of the Ohlone people. After years of being erased from the history books of this unique Presidio land base, the Ohlone are beginning to be recognized and honored as the original inhabitants of Crissy Field. The hope is that through more education, they will be acknowledged as having lived and worked throughout this new urban park.

Most of us find it is easier to separate ourselves from nature than to embrace the liquid mystery of our union with it. As freshwater disappears on the earth, so do the water stories that remind us that we too can freeze, melt, conceive, and evaporate. We too can construct a confluence of cultural rivulets where the natural and cultural coalesce.

### *Notes*

1. Edward H. Schafer, *The Divine Woman: Dragon Ladies and Rain Maidens* (San Francisco: North Point Press, 1973).

2. California Department of Water Resources, *California Water, Looking to the Future,* Bulletin 160–87 (Sacramento: California Department of Water Resources, 1987).

3. Andrew Neal Cohen, *An Introduction to the Ecology of the San Francisco Estuary* (Oakland: SF Estuary Project, 1990).

4. Delphine Hirasuna, Roger G. Kennedy, and Robert Glenn Ketchum, *Presidio Gateways: Views of a National Landmark at San Francisco's Golden Gate* (San Francisco: Chronicle Books, 1995), p. 2.

5. Kenny Ausubel, *Restoring the Earth* (Tiburon, Calif.: H. J. Kramer, 1997).

6. E. W. Gifford, "Indian Mound Found in Presidio's Limits. Bones Discovered: Relic May Be Preserved as Curiosity of World's Fair," *San Francisco Examiner,* August 15, 1912.

7. Martin Mayer of the GGNRA, NPS, to Kathy Nelson of the Northwest Information Center, California Archaeological Inventory, Sonoma State University, March 16, 1990.

8. Rodger Heglar to Lieutenant Gonzales of the Sixth Army Headquarters at the Presidio of San Francisco, April 23, 1973.

9. Archaeological Site Record, Department of Parks and Recreation, December 3, 1998, prepared by Holman & Association, San Francisco.

*In the Ice, 1879.* Courtesy, Peabody Essex Museum, Salem, Massachusetts

# Water Aphorisms

*Malcolm de Chazal*

*Translated by Irving Weiss*

Water meanders on a completely smooth surface and toboggans down the glossiness of leaves.

The feel of the neck of branches, of the mouth of the flower, of the belly of water, of the haunches of fruit. O leaves, your wet tongues.

The water that slides along oily surfaces of seaweed in full sunlight drags behind it the complementary pressure of light upon it, like horse and rider jerking together in full gallop.

The babbling brook struck by a headlong gust of wind chokes the sounds deep down into its throat until the liquid lips emit a human cry.

Water talks with its mouth full; the air with its mouth open. Which is why we can understand the language of the wind better than what the brook burbles.

All colors take to water in blue. A red rose on a blue dress ploughs through the weave of the material like a boat.

The wind is the wave's oars, the current is the scull. The whirlpool is the water sculling in a circle.

What you hear in the voice of the seashell is the waves whispering in the wind's ear.

Water leaps out of its body before throwing itself over the ledge of a falls, like a diver just before he actually dives. All naturally falling bodies fall out of themselves.

The wind that blows the rain around swings one of the rainbow's ends closer to the eye.

Wind overcomes water; water, granite. The grindstone has its way with steel. In the end feeling shapes thought.

Waves clashing on the ocean's expanse swell their hips and break out into wavelets of applause like a dancer shuddering with the drumbeat at the height of her efforts.

The thirsty animal drinks with his teeth, crazed with the desire to eat water. Man, too, sees his shrivelled lips turn to dust with the longing for water.

Absolute liquid would tremble perpetually. Absolute solid would be totally inert. In the first case the least movement would spill the liquid from its container; and in the second, by the effect of total inertia, the force of acceleration necessary to move the smallest object would so far surpass human effort that man would have to abandon his struggle for life and let himself die on the spot.

Still water is the prisoner of its container, running water is the prisoner of its source of movement. Water is free at last only in rain—when the wind keeps its distance—and in the toppling fall of the fountain. And so with us, who are never really free of women except at orgasm and the few moments afterward. Living makes us our own prisoners. Dead, we belong to others. The only real freedom, then, is to drown and sink briefly in ourselves.

Foam swims with its fingers. Airy-tentacled, foam is a cuttlefish on the surface of the sea. We all know what it is like to plunge our hands into foam, thousands of little feelers clutching us like the arms of an octopus. A giant spider of water lying in wait to embrace its liquid-fast prey, foam is the sweeper of streams and the brusher of lakes.

Foam is the most perfect of all swimmers. The body of foam is so perfectly relaxed that it lets itself be borne by the water without the effort of a slightest gesture. We would be better swimmers if our minds were disengaged. All current swimming records would be broken if we could only teach our athletes to unlearn thinking their movements.

Water was made to fall, run, rush out, be suspended or carried, or even leap right through itself—and at the other extreme tiptoe its dewy way along the surface of a leaf. But water's uneasy sense of balance gives it fits of terror and nausea. Giddiness overtakes the fountain at the height of its trajectory so that it never really regains consciousness until it touches ground. Water is the model of all those who try to live two lives at once: sooner or later they fall into a trance and regain their "balance" only when they "land."

The waves are always playing leapfrog with the frog disappearing just as the leaper hangs astraddle. Lines of breakers are a cavalry formation with the riders presumably in the troughs. The only rider of waves always in plain view is a raging wind. Light winds keep falling off the crests and remounting like a horseman's legs bouncing a rise and fall against the flanks of the galloping horse.

Flowing water is always dancing. Water leaps, glides, runs, and dances, bounding as an ocean wave, galloping headlong as a stormy breaker. It never merely walks. To walk you have to balance on one foot while planting a step with the other—but water is always too busy sending signals of movement all over its body at the same time. Water is a pair of compasses jumping, leaping, dancing, running, with both legs at once, never one leg fixed while the other one rovers. Water could never walk without faltering. It would always threaten to break its neck at every step like an insecure toddler. If water had to learn to walk first, it would never have managed to get on its feet at all.

If we could only take apart the sounds of water by bellying out our ears to catch every bit of babble separately, it would be like opening up the blossom of its sound petal by petal instead of straining to hear the cries gushing out of its depths. Every little lip murmur would be audible in the chatter of water no longer swallowing its words. Think of the difference: syllables instead of gutturals, liquid words emerging at last, perhaps even complete watery sentences. All ears and alert, we would begin to make sense of water's voice instead of groping for its cries. To extend our hearing in this way would of course mean to spiritualize our ears with the psychic prerogative of the angels.

There's always something left over in the sound of water falling: falling water sounds like final chords that will never end. The sound of water falling is the prolonged agony of sound refusing to give up its last breath, a death rattle reborn with each last gasp, the delirium tremens of overindulgence in sounds, a voice gargling on a mouthful of water, like chronic huskiness rattling in words or like the letters rattling inside their own words.

## Rain

*Joseph Bruchac*

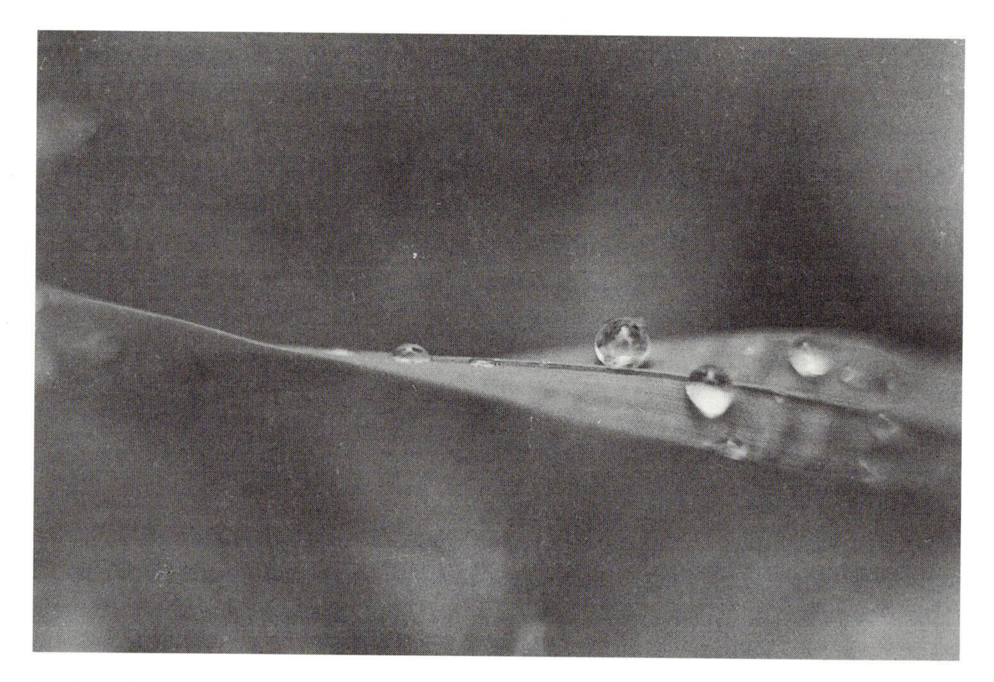

Matthew Ebinger

First silence
and then
almost like a sigh
the gentle
harp note plink
of a raindrop
that fell from so high
we can barely imagine
its sparkling journey
down
  from
    the
      sky.

It is the beginning,
the very first
note of a melody
older than breathing.

Long before we
who walk, swim or fly
arrived
this pond was singing.

## Watershed Governance: Checklists to Encourage Respect for Waterflows and People

*Peter Warshall*

*The rain falls on the just*
*and the unjust fella,*
*But mainly on the just 'cause*
*the unjust got the just's umbrella.*
—African American tapdance rhyme-song

For watershed governance, you need the discipline of working rules and a good sense of humor. Admire humans and their leaky canteen-like bodies, and gently, firmly cover their greedy mouths before an insatiable thirst destroys the town. For where there are enough humans and not enough water, hydro-squabbles spew forth. Mark Twain, as usual: "Whiskey's for drinking; water's for fighting."

I served as an elected director of a utilities district for eight years, which meant endless hours wondering how local humans fit into watersheds. During this period and a subsequent twenty years as consultant, especially in sub-Saharan Africa, I can't think of any aspect of waterflow that citizens didn't douse me with: dam building or dam removal; water wrung from clouds or water pumped from depths; water exclusively for humans or water exclusively for birds; disgustingly dirty water; bottled water of questionable purity or flush toilets of questionable efficiency. It's all been important: a lake to preserve the reflection of the moon; cattle slobber in the feedlot; the unexplained healing powers of water in a steam bath or swimming pool.

I lucked out as a politician and, on a local level, helped build a totally recycling sewage system with prodigious artificial wetlands; encouraged self-sufficient home-site sewage systems; and ensured pure drinking water. Political promises—clean and cheap water, an adequate supply for community desires, and a fair share for all (farmers, householders, bird watchers, and retailers)—miraculously worked out. But elected office taught that water governs town life more profoundly than do cheap and clean supplies. Waterflows, for instance, governed housing densities and town growth patterns. Citizens laughed cynically at zoning rules, which appeared to change with each election. In their eyes, planning commissions and their representatives were fickle (if not sophisticated in a corrupt manner). But a small-diameter pipe that limits waterflow limited water supply and guaranteed limited sprawl. Waterflow shaped the town. Citizens refused to approve bonds for bigger dams to increase supply or wider pipes to deliver more water. They saddled the town government with a fourteen-year law fight. But the cost of litigation appeared more desirable than the cost of new waterworks and housing growth, so citizens insisted that water and truth be tightly tied. Truth-as-repetition-truth could not sway them. They did not believe that the more an authority said something, the more it must be true. They did not believe: zoning managed housing growth and land use patterning. The truth they came to was truth-as-it-always-is-truth. Waterworks shape the lives of watershed communities.

It was a time when this kind of skepticism sprang up in many communities. I remember Nassau County, Long Island, citizens had been told by media and experts that their poor-quality drinking water stemmed from too many failing septic tanks. It was repeated endlessly with the normal parading of experts. The citizens voted millions to replace the septics with a modern sewage treatment plant, only to discover that, after sewering allowed greater housing densities, the poor water quality came from pet dogs that pooped in local flood control basins. Doggie scoops would have solved the problem.

Like myself, each citizen I've talked with—about voting, protesting, lobbying, or occasionally thanking elected representatives—has at one time or another turned inward to reflect on his or her feelings and intentions about water. Their images and memories of waterflow have applied a kind of aqueous moral pressure, an insistence within their selves to act, making a choice for the human community and, by extension, the watershed. The package of checklists that follows, almost one-liners, became important to me in considering water, this moral pressure, and power.

Out wriggles the first salient truth: *water creates an inescapable contract.* We cannot choose to be part of its flows. Our bodily plumbing is always linked to the biosphere's. We may live upstream or downstream; right bank or left bank; drink surface water or groundwater; seed clouds or fish in a river; wash city streets or feed a cooling system to regulate the temperature in a nuclear power plant. In all actions, we never free ourselves. Water is biospheric life support married to the laws of thermodynamics with no divorce clause. This inescapable hydro-contract leads to an inescapable social contract: *we need working rules to maintain hydro-harmonies.*

### *The Nature of Water and Watershed Governing*

Consider how water governs itself. Bemused awareness of water's "self-governance" must always be first on the agenda because waterflows have a mischievous way of subverting and toying with our species's meager abilities at foresight. Deny or fight waterflow's always-true-to-itself nature, and pesky water nymphs stir up trouble. The Romans were so impressed that they considered free-flowing water a wild animal. Only captured water—like a caged animal—was subject to law.

Here are a few personality traits of water's wild animus:

- Water is ethically neutral.
- Water has its extreme moments.
- Water is bulky, erratic, mobile, evasive.
- Water transforms without permission, assimilating organic and inorganic chemicals into itself more than any other substance on the planet.

Water is neutral. It is indifferent. Waterflows don't care if they are polluted or clean, flooding or trickling; in the president's semen or the sweat of the pope's benediction. Water is an anarchist. Water blithely ignores property boundaries on land, underground, and at sea or in the atmosphere. Fickle water nymphs confound the best and worst of serious ethics. A slave in Missouri watched a flood alter the river's course, and woke up a freeman in Kansas. Because of water's neutrality, human rules about access to and use of waterflows must be very clear. Not for water's sake—water will do what it does—but for human comfort and the sake of other living creatures that human water management influences.

Water has its extreme moments. Droughts, floods, blizzards, and turned-over lake bottoms cause obvious suffering and pain to humans and other planetary creatures. Yet hydro-extremes must be considered normal in all watersheds and in all hydro-compacts. Buffering or reducing the intensity and frequency of hydro-extremes is a modern human obsession. We don't like waterflow's dramatics. We just won't tolerate them. Waterworks to tame waterflow is perhaps the most capital-intensive, property-contentious, and socioeconomic inequitable aspect of watershed management. Water could care less.

Water is erratic, mobile, and evasive. Even without extremes, water has a maddeningly contrary nature. On the mid-reach of the White Nile, the Dinka teach this aphorism:

> Spring rains in a dry spell.
> Drops club ants on the head.
> And the ants say: It's god's work.
> And they wonder whether it helps.
> And they wonder whether it injures.

Waterflows, for instance, disappear as groundwater, or evaporate as vapors just when you need them, and when you least expect it. Rivers change course despite billions of dollars of flood-control channeling. Much of modern water politics—its hope and optimism—starts with a denial of aqueous unruliness. In the United States especially, we are a very "solution"-oriented culture, yet water sticks its tongue out at our idealism and pragmatic pride. Sign a contract for long-term water delivery and the rainfall pattern mutates, rendering the contract irrelevant (Colorado River).

Water is equally unruly because it transforms quality without permission, assimilates organic and inorganic chemicals into itself more than any other substance on the planet. Water is the great absorber, adsorber, and biochemical renovator. Sometimes too salty or muddy or rich in nutrient. Sometimes too radioactive or brimming full of the emotions of local history. Count on drinking from a well, and then find strange toxic brews showing up from unexpected sources (Tucson groundwater). Farm for maximum corn yields and discover that the runoff has killed off the fishing industry (lower Mississippi). In watershed governing, the price of water is not water itself, but increasingly the price to maintain its beneficial qualities. Drinking water and wastewater treatment and desalinization are governance issues because every gallon needing purification needs energy to purify it, and energy has a cost.

Water is bulky and heavy. In the United States, the weight of deliberately transported water exceeds the weight of all freight moved by railroads, barges, and trucks. The price of water is, in many river basins, the price to move it. Governance, for instance, now includes moving water hundreds of miles from the Colorado River to five major jurisdictions (California, Nevada, Utah, Arizona, Colorado, Mexico). Pumping this much water over mountain ranges in an open aqueduct that concentrates the salts requires energy for both lifting and purification. The infamous aphorism of California—"Water flows downhill, except when it flows uphill toward money"—puts the unholy triumvirate of river basin governance into one sentence. Waterflow, the energy flow to move it and purify it, and the cash flow to pay for it all spring from water's bulky weight, its resistance to cheap transport as a liquid, and its perverse acceptance of what life dumps into it.

### *The Human Life of Water*

If water itself were not such a pain to deal with, then stirring in human feelings about water will surely drive watershed citizens, leaders, and managers to drink. Elected officials, planners, engineers, fishers, wild rice gatherers, duck hunters, wetland birders, swimmers, water skiers, riparian restorers, barge captains, wastewater treatment plant operators, canal and irrigation ditch workers, farmers, fire managers, plumbers, homeowners, and others can all lose track of hydro-essentials: that respecting water with all its contrary and mischievous behavior is the touchstone of a kind politics.

Here's the second course of priority aphorisms for those who dare mix water-thought and a big heart for their human communities:

- Water is life.
- Water is precious—a strategic resource in the maintenance of social cohesion.

- Water is deeply embedded in watershed landscapes, is tenaciously site specific.
- Water is more than a commodity with production value.

Water is life. Like liberty and the pursuit of happiness, minimal hydro-security is a basic human right. A healthy water supply supersedes all other economic and legal dictates. "Our bodies are but molded water" (Novalis). When the minimum vanishes (as in the Sahel after sixteen years of drought), all sense of compassion toward the landscape, other humans, or nonhuman creatures fades. Hydro-security is a two- to three-day proposition. Without water, thieves, beggars, migrants, and riots emerge in the dust. We forget. We are lucky; we forget. Water first! This essay assumes watersheds with a basic water supply. To give proper attention to crisis watersheds and human coping strategies, it's just another endeavor.

If water is essentially the blood of our lives, it is also the blood of non-human species and productive habitats. Although honoring biospheric life support is not common to the majority of humans, there are early indicators that the pursuit of human happiness in the future will incorporate a demand for water for flora and fauna. Without much of a stretch, water can be seen as the basic right of all living creatures.

Water is precious—a strategic resource in the maintenance of social cohesion. "'Til taught by pain, man knows not water's worth" wrote Byron as he traveled into the more arid lands of Greece and then Turkey. Deserts and droughts were stark teachers. In Morocco, death can be the penalty for wrongful use of an oasis or allowing nonassigned goats to eat leaves from wadi trees. In the contemporary Jordan/Syria/Israel river basin or Ethiopia/Sudan/Egypt basin of the Blue Nile, hydro-balances are delicate, and water sings a military tune. Destroying waterworks is the ultimate card to be played in the poker game of diplomacy and war.

Even in times of peace, every recipe for every good and service of value to human societies says: Add water. All energy (coal, nuclear, hydro, wind, diesel, geothermal) requires water. Waterflow drives the planet's green machine, churning out food and entering all product development from computers to underwear. It provides services for free: modifying climate, aerating and purifying waterflow as it rushes down rivers, biochemically renovating pollutants, providing a medium for fish to grow. Agricultural, industrial, and information-based societies require incredible waterworks. From the Fertile Crescent to the

Yangtze of China, hydraulic civilizations have come and gone with the ability of their governors to keep water flowing to communities without unexpected harms.

Water is more than a commodity. Water bodies are great sources of joy. Sloshing around in tanks and ox-bows, swimming pools, and creeks nurtures giggles; sitting by the ocean quiets the soul; and steam baths and jacuzzis hold unexplained healing powers of water. In our dreams, reflections, and stories, water as perhaps no other substance gushes with beauty, spirit, grace, and stories of long-term community life.

In watershed governance, water's production and consumption pricing is only part of the picture—sometimes a very small part. Purely economic models will never suffice. For instance, economics can play second or third fiddle to cultural stubbornness. American love of golf has, in many law cases, insisted that it is a higher and better use than an in-stream flow for an endangered trout. Golf is an American ritual, even a rite of passage. Or many Hindus insist that bathing in the Ganges is healing despite their exposure to water-borne disease.

Nonconsumptive uses like leaving enough water in a river for swimming and fish conveyance have indirect, difficult-to-assess, economic benefits. They drive politicians who like cost-benefit analyses bananas. Yet all watershed governance must include allocations that may have no apparent financial value, such as enough water to preserve the simple reflective beauty of lakes. In the parlance, hydro-beauty has a nonmonetary existence value . . .

> That flowing water! That flowing water!
> My mind wanders across it.
> That broad water! That flowing water!
> My mind wanders across it.
> That old age water! That flowing water!
> My mind wanders across it.
> —Navajo Mountain Chant myth

As opposed to other flows (cash, information bits, or energy), waterflows are hard to model, stylize, or generalize. To the great frustration of the human dreams and reason, waterflows are maddeningly site specific, embedded in unique watershed landscapes. The ways that watersheds gather, store, channel, schedule, disperse, concentrate, dilute, and biochemically transform water quality depend on details—the specific weather, the specific configuration of the aquifer, the specific flood pulse and channel width, the specific mass loading and concentrations of chemical ingredients. Where I live in Tucson, salts and pollutants concentrate as they descend downstream because of water losses from high evaporation. In my mom's watershed (the Hudson), salts and pollutants dilute with each downstream tributary. In the more formal language, watersheds depend on situational variables more than aggregate or summary variables. Sit pleased as Punch with an aqueous generalization, ignore the individuality of a watershed's persona, and the waterflows will play you for a sucker.

To harmonize watersheds and their human communities, Watershed Mind learns to be humble, intellectually insistent, and soulfully patient. It is hard work, and the uncertainty principle thrives in water. How many times have engineers spoken with solemn precision about a creek's design flow and with equal solemnity about its water quality, convincing a utilities board of its grasp on reality, and how many times has some other aspect of water's contrary nature or watersheds' surprising adjustments eluded our mental picture?

Governing bodies spend much of their time dealing with the unintended consequences of earlier "solutions" to their watershed's management problems.

### *Watershed Rivalries*

*Rival* comes from the Latin word *rivalis,* meaning river. In the Hollywood version, two groups of cavemen brandishing stick spears scream and prance on opposite banks. The river is the rule, the no trespass sign. But outside Hollywood, watershed rivalries come in many more flavors. Sometimes people fight over access to water sources; sometimes they agree on access by fight over use privileges (how much and when it can be taken); sometimes they fight over a damaged or degraded supply. Sometimes, as with the Taos pueblo fight over sacred Blue Lake, they fight to keep a body of water untouched.

For aqueous contemplation, here are a few simplified post-caveman rivalries:

- A sewage plant treats the wastewater from a town and then sells the treated wastewater to a golf course. It uses the revenue to increase the size of the treatment plant, anticipating new customers. The residents protest. They claim the wastewater was theirs, and the profits from selling the water to the golf course should go toward reducing their monthly fees. A conservationist group also protests, claiming that the treated wastewater (discharged into a creek) had supported a wetland bird sanctuary and, once dedicated to the creek, was no longer the property of the wastewater treatment plant. It should neither be sent to the golf course nor used to reduce sewage fees. These cultural, environmental, investment, history, and water reuse confusions are typical. How to navigate?

- A water company has impounded a stream for hydroelectric production and releases water only when power is needed. This causes the fish population to crash. Trout fishermen sue and demand a more even release, no matter what the demand for power. The company says that the fishermen can have the water if they pay the equivalent of lost revenues. The fishermen say that the water rightfully belonged to the stream first. If the company refuses to release waterflow to help maintain the fish, the hydropower company should pay damages. Who wins?
- A farmer has spent time and money installing water conservation equipment and has cut his water use by 35 percent. But he has no additional land, so he now leaves 35 percent more water in the river. He claims that this is his water (by previous use) and wants the downstream farmer who has more land to pay him for it. The downstream farmer refuses because the upstream farmer has not "used" the water left in the river or transported it to him by pipe. The upstream farmer lost his property right for being a good citizen.
- A series of farmers pump out groundwater, lowering the level of a spring so it no longer flows. A Native American group claims the spring is sacred, a crucial element of their healing ceremonies and spiritual needs, and part of their cultural heritage. Because the spring has no "development" associated with it, the law states that it is not the "property" of the tribe. Unless the Indians can demonstrate prior use of a specified volume of flow, they cannot claim the spring. There are no written documents (only oral stories) supporting prior use, and the volume consumed (compared to the outflow desired) is minuscule.
- An upstream sewage plant treats millions of gallons of urban water to the highest standards set by the government. This high-quality effluent enters a saltwater estuary and causes osmotic shock (the creatures that

thrive best in salty water are harmed by the influx of freshwater). A conservationist group sues to stop all discharges that might hurt an endangered fish, even though the treatment plant has performed better than all discharge requirements.

In a newly settled watershed, rivalries center on capturing and dividing unclaimed supplies and capital to build the waterworks. In industrialized nations, the waterworks have already been built, and most water rivalries start when citizens and governments change water from agriculture to urban use; transfer waterflow between watersheds; divert more water from in-stream to off-stream or underground to aboveground uses; degrade water quality and damage downstream needs; subvert taxpayer senses of equity; negotiate the damages from acid rain; or size sewage and water pipes that influence housing growth.

In all rivalries, healing watersheds means healing the angers, ignorance, and desires, if not bitterness, of watershed citizens. It's an old story, told in China centuries before the common era:

> The sage's transformation of the world arises from solving the problem of water. If water is united, the human heart will be corrected. If water is pure and clean, the heart of the people will readily be unified and desirous of cleanliness. Even when the citizenry's heart is changed, their conduct will not be depraved. So the sage's government does not consist of talking to people and persuading them, family by family. The pivot (of work) is water. (Lao-tzu)

A postmodern Lao-tzu might carry a crib sheet to help dive to the heart of the matter:

- How much unclaimed or overclaimed water resides in the watershed?
- Are there scarcities by location, season, or effective soil moisture and runoff?
- Which or whose vested rights predominate?
- Where are the watery places of special nature? What aquatic nonhuman creatures require protection? How much water is needed for nonconsumptive uses?
- How much of the flow resides in the watershed's natural ecostructures (its channels, wetlands, soils) and how much in the human-built infrastructure (its pipes, aqueducts, storm sewers)?

- What unconrollable "externalities" (such as leaking groundwater, cloud seeding, climate change, interwatershed transfers) impinge on workable hydro-rules?

### *Institutions and More Institutions*

Lao-tzu's great contribution to watershed governance was this: *Always give priority to water over human special interests.* No matter how charming human ideals, poetics, political rhetoric, divine revelations, promises, or factoids, hydrophilia is the best consensus builder. I always check out a rival or ally to discover if he or she actually likes water, finds it both amusing and profound, knows water has its own way of being. If the person did not cherish water, he or she was probably the all-too-common shuck-and-jive citizen-leaders who thrive on the hydro-politics game (but not water-watching). Ultimately, no matter which side of the hydro-squabble, they do not have the skills and affection for water to overcome special interests. The rarity of hydrophilia can be appreciated. Not for another 2,000 years, with the birth of John Wesley Powell, did another human write as well as Lao-tzu about hydro-communities and the skillful means to build hydro-harmonies.

To achieve peaceful resolutions of water rivalries, Lao-tzu would eventually need to gather the community into a forum. Within these human "confluences," the community's working rules for waterflows emerge. *Institutions*—broadly, the working rules agreed on by a community—*carry sustainability.* The intensely anthropocentric task of creating rules for human behavior is the heart and soul of watershed governance. In nations where institutions are weak, humans commonly kill each other over water deals. Where watershed institutions are unfair or hopelessly outdated, disenfranchised and angry citizens gravitate toward cheating on water allocations and stealing waterflow (especially at night). The Umatilla basin in Washington is a classic example. If pushed "too far," embittered water users monkey-wrench, obstruct by nonviolence, or attack. (The last U.S. militarized water war was in 1934 over the Colorado River. No one was killed.) Uncomfortable with institutions that coerce humans to accept dictated waterworks (the Three Gorges Dam in China) and imperially dictate waterflow rules, I have spent much of my watershed life trying to destroy watershed governance rules and institutions creatively and replace them with more responsive working rules.

Citizens discover new rules by gossiping in the local café, walking the watershed, questioning authorities, adjudicating, arbitrating, mediating, casting their votes, and negotiating. Governments institutionalize their watershed rules in myriad ways: informal agreements, hiring a water master to arbitrate flows, electing utilities district regulators, forming a private corporation or a public water or sewer district or a river basin quasi-public task force, relying on state water departments, depending on federal agencies, and so on.

In a New Mexican watershed I recently visited, the state's rules didn't work. It was nearly impossible to deliver water to the priority rights holders and then block up the irrigation canal for the next in priority (who was upstream), then reopen the floodgates for the next in historical priority. Water was lost, the process took too much time, and it was too complex to remember. Quietly the watershed residents forgot the priority sections of the water law (keeping the volumes intact) and simply trusted the water master to let water out of the ditch in a fair and equitable manner. He had done so for over thirty years.

Local citizens have experimented with their own forms of shadow governments, essentially creating instituions that do the job that the government or private utility should have been doing. Sometimes, when strong enough, members of these parallel institutions challenge competing special interest groups or hydro-bureaucracies. They move from shadows to daylight. That's how I was elected.

In the early 1970s, for instance, no working rules guaranteed waterflow inside the river channels of U.S. watersheds. The salmon in the Columbia River basin and their human voices (fishers and Indians) could only tell the tale. Slowly the fish and their human spokespeople gained legitimacy among the hydroelectric boys and the irrigators. They had to fight to join the small, officious clique that dreamed up all the rules. They found themselves trapped by the past (impassable dams), finances (the costs of fish ladders), and the old habits of the bureaucrats. What is fascinating about their now limited success is how much rested on their ethical arguments about fairness to fishers, treaty signatories, and the fish themselves, and their intimacy with the actual riverflows and fish runs. In Lao-tzu style, the power of intimacy with their watershed overcame the power of money. When the dream that it might be possible to rehabilitate past damages took hold, the shadow governors began to reshape the rules. Removing dams is now considered a reasonable alternative.

For those who are starting their first watershed shadow government, here, in the lineage of Mr. Powell and Elinor Ostrom (in her brilliant *Governing the Commons,* 1990), is a practical checklist. It's long but it can save a lot of community heartache. If I've forgotten or missed an item, send it along for this essay's next revision.

*A good watershed institution:*

- Opens its doors to public participation and ensures that everyone has equal access to information.
- Gives a priority to local watershed needs (the watershed of origin).
- Has a well-defined degree of local autonomy so it can peacefully custom-design rules and change them, despite regional water budgets and ever-fluxing values of the culture.
- Is accountable, competent, respects its own laws, adapts laws to water-years, finds the best local incentives and disincentives, and protects human rights.
- Allocates a certain volume of flow to the private (competitive) sector, a certain amount to long-term stability (usually publicly enforced water rights), and a certain amount for existence value and basic human needs (untouchable by either the public or private sector).
- Publishes the transaction costs of the institution.
- Informs and asks watershed citizens about the harms and costs to downstream communities, slope stability, soil capital, channel equilibrium, water qualities, and the flora and fauna within the watershed.
- Informs citizens of the financial costs and benefits of each waterflow usage as a segment of the total cash flow of all watershed activities. Are there important multiplier effects? Is there any bonding capacity left?

*Good watershed governors:*

- Decide who is eligible to make decisions about watershed waterflows.
- Decide what actions are allowed or constrained, forbidden, required, permitted, and encouraged.
- Set inflexible rules for basic water rights for both humans and instream flow. Set very flexible rules to preserve an arena of competition based on allocations for the highest price per gallon.
- Decide what aggregation rules will be used (the series of steps required to protect, challenge, and change the rules).

- Decide what water-year procedures must be followed (emphasize incentives).
- Decide what knowledge (information and wisdom) must or must not be provided.
- Decide what payoffs and punishments will be assigned to individuals depending on their actions.

### *The Dr. Watershed Kit for Essential Watershed Governance*

*All things are dissolved by fire and glued together by water.*
—Plato's follower

Perhaps this essay contains too many lists: one for water's nature, one for human communities and water, one for good watershed governance, one for starting a shadow government. Consider them as points in a flowing river—pools and riffles, eddies, and the mainstream tongue. You can come back to them and sit there water watching as needed. I have one more list, a short one of tools that have helped me resolve water disputes.

*The Water-Year*
My favorite tool for promoting peace, flexibility and harmony is the water-year because in any particular year of rainfall, only certain actions are possible and certain sacrifices must occur. This is the humbleness that water teaches. I learned about water-years in California. The state, by opening and closing dams, allows different volumes of flow into San Francisco Bay in any of five types of water-year: normal, dry, critically dry, wet, and super-wet. While a myriad grumblings occur, everyone basically understands that you can't apply the same rules to different volumes of river flow. By framing governing guidelines by water-years, watersheds as different as industrialized San Francisco Bay or nonindustrialized Lake Chad have begun to create flexible rules for community hydro-design.

*Water Efficiencies*
The way for politicians to resolve water issues is to stretch the supply. They love consultants who come by and say, "We can give everyone the water they need." So to ensure fair distribution and equity in watershed governance, all working rules consider more efficient use of the supply: promoting water-saving devices and closing loops by recycling (more with less), optimizing the

use of energy (less water lost as a coolant and feeder stock in utilities and industry), optimizing the use of material flows (less water lost as an ingredient and medium for catalysis in chemical and food industries), minimizing waste generation (less water lost as a dilutor of harmful residues), and maximizing cascading uses (water of varying qualities assigned to their most appropriate use). Efficiency by itself is nowhere near the complete picture. Cultural needs may be inefficient. But appropriate technology, from drip irrigation to adopting the least-water-intensive industrial process, is a crucial tool.

*Incentives*

Watershed accounting looks at each cash flow and waterflow and decides if the money spent can be justified. This conceptual tool looks at the kind of temptation or incentive that money provides: perverse, harmonious, or punitive. A perverse incentive is, for instance, subsidizing the energy cost of pumping water from an aquifer faster than it can be replenished, or paying to rebuild a town in a floodplain when moving the whole town inland is cheaper. A true incentive is allowing a farmer to sell any water from his vested right that he conserves through changing his irrigation practices. A disincentive is financially penalizing or closing down a polluter that degrades water quality used for drinking. Listing rules within these three categories forces healing into the practical realm. First and foremost, rid the watershed community of perverse incentives. Second, encourage and find new incentives. Third, restrict disincentives to violations of the clear rules of access, use, reuse and "security of the minimum."

*Allaying Fears*

What is "clean" water? What is "adequate" supply? What is "harm" or "damage" from pollution, acid rain, or upstream watershed alterations that cause downstream floods? Judging risk is not easy. It's hard to remain skeptical of those who claim it's all fine (don't worry) but not cynical. Does the bacteria enterococcus cause sickness in saltwater, when the studies have only been done in freshwater lakes? How much enterococcus is dangerous? Given imperfect knowledge, how much to pay for what technology to reduce enterococcus along the shoreline? Can we allow one infection for every thousand swimmers? One in ten thousand? But my kids are on the beach! Is the politico exploiting local fears so he can appear as the white knight who will save the children? Or has he made a deal with the engineering firms that will profit from "solving" what may be a nonproblem? Or is the health department in need of more federal funds to keep the bureaucracy humming, and enterococcus is a way to drum up financing? Or is the department doing its duty to protect the public health? Are the media looking for dramatics? Is the issue smoke-and-mirrors, dueling experts, talking heads, or a real risk? And who really understands the "risk assessment" models of successive approximation, correlation and regression, extreme or central values that allegedly prove safety?

Once again, the truth issue brings citizens to shadow governments and skillful leaders who respect water first and foremost and spend time understanding its wily ways. In my experience, intimacy with the truth of water rarely ensures success in the political arena. But a combination of empirical hydro-truths and a sense of fairness can hold sway. For instance, the environmental justice movement, a movement among minorities whose watersheds became the dumping grounds for water-borne pollutants, gained strength by combining meticulous water quality data, epidemiology, and ethical arguments for civil rights. There are dozens of similar stories from the *Clearwater* yacht on the Hudson that took water samples to Love Canal moms.

*Rules for Changing Rules*

In Africa, many citizens told me: "We like America. When you change leaders, no one gets killed." The last tool rests on wisdom, a deep wisdom. How to educate leaders to change rules without bloodshed? How to change rules without installing the internal desire for revenge? Watershed governance falters when there are no rules for changing rules. Modifying the ownership of water, communal holding rights, and the access and time of use rules are essential conflict-resolution arenas for sustainable watersheds. When the existing rules are obsolete, negotiations are tedious and ugly. The issues are pushed aside, and I've heard "We'll deal with that later." But rules to change rules are as important as knowing that cutting trees will cause floods in the future. A tool to remember.

## Water Watchers at the Holy Spring

> *Let the most absent-minded of men be plunged into the deepest reveries . . . and he will infallibly lead you to water, if water there be in that region. . . . meditation and water are wedded forever.*
>
> —Herman Melville

There are certain watery places where understandings and healing are easier. Springs are classic. Here, the essence of watershed governance delightfully displays itself. The spring is a specific place; watershed governance is always in unique geographies. The spring is a literal source of life. So is the ability of humans to govern themselves with kindness. The spring, coming from underground, makes one contemplate the unseen and unknown; major skills in the wily politics of waterflows and human desires. Springs tend to have special waters, healing waters, and healing plants; humans like to gather at them and make special trips or pilgrimages to them. Springs and community governance carry unique history ("Remember what happened at the spring in 1876?"). Even when not physically at the spring, springs become images of life-giving, reverence, and contentions that encourage humans to meditate on their dreams and what they will accept in life. Springs reinforce the value of water watching, of placing water first in the resolution of differences. Springs bring humans back to basics.

The best hydro-citizens are water watchers. They love to discuss the rainfall or the meanderings of their river. They love to bicker about who's sucking up how much water at what price or how many parts per billion are dangerous. Without maniacal water watching and daily palaver, water and governance have a hard time mixing. The best water watchers combine hydro-contemplation with water's teachings: all governance must rest on interwoven life support; qualities are as important as quantities; flexibility is essential; changing the direction of flow requires tremendous energy; and all creatures are equal in the kingdom of waterflow. Michelet, an eighteenth-century utopian, understood water's ability to mix what humans might want kept separate. He wrote: "Religion, education, government. These are the same word." And, then, perhaps, he took a hot bath.

## Tidal

*Jody Gladding*

But it's not light
*you're* carrying
said the moon

it's water
the live weight of water
a basin you lift slowly

shift suddenly
and it's water's wave
that rocks you

when it spills
not the way light spills
a rush of water

all will and habit
finds a new course and takes
everything down with it

to leave you
to collect yourself
beside a shimmering

new body
of water—so calm
in its new bed

so light.

# II

## *Teachings of the Flow*

# In Salmon's Water

*Freeman House*

> *Sometimes your storyline is the only line you have to Earth.*
> —Sharon Doubiago

I am alone in a sixteen-foot trailer by the side of a river. It is New Year's Eve, 1982. The door to the banged-up rig stands open, and when the radio is off, I can hear water in the river splashing endlessly over cobbles. The oven is on full blast. Its door hangs open too. The heat rises to the ceiling in layers, ending at the level of my chest. My face is hot, but my ankles and knees are cold and damp. On the radio the Grateful Dead and fifteen thousand celebrants woozily greet the new year at the Oakland Coliseum. Ken Nordine's deep beatnik baritone drones on. Ken Kesey babbles. Any moment now, Bill Graham, undressed as Baby Time, will be lowered from the rafters. The band lurches through the music, loses the thread entirely, and after a long time finds it again, the beat loose and insouciant throughout. The band seems to say, "See? Told you we could find it again." It all makes sense with enough LSD, I suppose, and I have sometimes lived my life as the Grateful Dead plays its music, drifting in and out of the right way to be, risking everything on an exploratory riff. But tonight I am focused and full of purpose. My only drug is a poorboy of red port, which I sip cautiously.

I turn the radio off and listen. Then, to hear better, I turn the lights off too. I am listening to the water. If you listen carelessly, the water in a rushing river sounds like a single thing with a great fullness about it. But when you begin to try to sort out the sound of one thing *within* the sound of the water, the moving water breaks into a thousand different sounds, some of which are in the water and some of which are in your mind. Individual boulders rolling along the bottom. The Beatles singing *ya-na-na-na*. The one sound breaks itself into separate strands that intertwine with each other like threads in a twisted rope. Some strands are abandoned as new ones are introduced, making a strange and hypnotic music. Listening to running water is a quick route to voluntary hallucination.

Among the many voices of the water, I am trying to distinguish the sound of a king salmon struggling upstream. It is a foolish undertaking and it never works. I hear a hundred fish for every one that is actually there, and then miss the one that is. The only sure way to locate a fish in this realm of sensation is to walk to the river's edge and play your light along the surface of the water where it passes through the weir. The king salmon may be large or small; it may weigh three pounds or thirty. If it has swum into the pen above the weir, I will pull the long latchstring that releases the gate that closes the mouth of the weir, so the fish can go neither upstream nor down. This doesn't happen very often in 1982.

A little more than three years ago, a state fisheries biologist told us that this race of native king salmon is done for. I am still not totally sure he wasn't right. The state Department of Fish and Game is spread thin. They can't afford to expend their scarce resources on a river that has next to no hope of continuing to produce marketable salmon for a diminishing fishing fleet. But a small number of residents of the remote little valley have not been able to bring themselves to stand by and watch while one more race of salmon disappears, especially the one in the river that runs through their lives. They have begun with little idea of what can be done. They've talked to other people like themselves, and also to ranchers, loggers, academic biologists, and commercial fishers. They have read books and sent away for obscure technical papers. They've developed a scheme that they hope will enhance the success of the spawning of the wild fish. Through stubborn persistence they've convinced the state to let them have a go at it.

By the last night of 1982, this little group has grown into a cohort of several dozen residents who are spending a great deal of time trying to forge a new sort of relationship to the living processes of their home place. We also have learned to deal with bureaucracies outside that place, and we have incorporated as the Mattole Watershed Salmon Support Group. We have raised money. We have entered into contracts. We are inventing our strategies as we go along.

I am part of that cohort. I am tending a weir with an enclosed pen behind it that is meant to capture wild salmon in order to fertilize and incubate their eggs. I am working by myself, which is unusual. Normally a crew of two or three would share these long nights. Most often, David and/or Gary, two of the people who initiated the effort, would be here. But it's a holiday. Everyone else has pressing engagements. The fish, however, know nothing of holidays. The spawning season is almost over, and we few who care for the salmon haven't come anywhere close to reaching the goals we have set for ourselves this year.

The weir looks like fish weirs have always looked on this coast—a fence angled upstream across the river from either bank at enough of a bias against the current so that it will not offer more resistance than it can endure. It closes off passage upstream except through a one-foot opening at its apex. In earlier times, a fisher with a net or spear might have stood behind or above the opening. For our purposes, the opening serves as the doorway to a trap, or to a pen. Although built from materials manufactured elsewhere, it has a funky look; it blends in. Panels of redwood one-by-one, grape stakes in another life, are spaced at one-inch intervals horizontally and lashed to metal fence posts pounded into the river bottom. Each panel has a chickenwire apron attached at its bottom. The aprons are held to the bottom by sandbags, gravelbags really, each one weighing about forty pounds. Filling and hauling the bags two at a time takes up most of the two-hour drill required for three or four people to install the temporary structure.

The salmon's progress upstream is one of many marvels of the salmonid life cycle. The grace and strength required to overcome waterfalls and other blockages, the stamina to endure floodwaters, the systematic persistence necessary to thread the maze that a big logjam presents—these are attributes so wondrous that we must consider them in the same realm as the mysterious intelligence that allows the creature to distinguish between the smell of her particular natal stream and the smell of the rest of the world of water. But when the fish swims into an enclosure that requires her to seek an exit *downstream,* she becomes slow and seemingly confused. It will usually take her some hours to discover the downstream exit that she found so quickly before, when it was the passage upstream. Her slow meanders seem now to lack purpose; escape from the trap, when it comes, seems almost accidental. It is as if nothing matters now that the path to the spawning gravels is blocked.

I had argued with my coworkers that we should take advantage of this weakness. We humans have little enough advantage dealing with such a marvelously functional aquatic creature, and I am a person who loves his sleep. Salmon have yet to recognize that we are trying to help them; they continue to evade us. We are social workers whose clients decline to be served. Use our terrestrial, linear intelligence, I said, to fashion traps that would hold the fish until morning. Wait to handle them until after a second cup of coffee.[1] And we had, for two years, fashioned beautiful traps to stand at the mouth of the weir. The traps had been built from the same grape stakes as the weir panels, and they had cleverly hinged plywood covers opening out from either side of the top. A three-quarter-inch cable slung all the way across the river from the top of the gorge at either side allowed a running block to be installed. Another line running through the block attached the traps to a hand-operated winch for installing the heavy hulks of the things in their exact locations, or for pulling them out quickly when the level and velocity of the rising water threatened to tear them apart or sweep them away.

But there was something about the traps—the sound that the waters made passing through so much enclosure, or perhaps the shadow that the things cast in the liquid boil below—that seemed to prevent the fish from entering. We had observed fish moving at dusk work their way right up to the mouth of a trap and then, in an instant, turn and disappear downstream. When they did enter and stayed for the night, they leaped against the plywood covers looking for a way out, wounding themselves and threatening their precious manifest of unfertilized eggs. Such a trap was too obviously a construct in service of human comfort, and we were, after all, seeking to serve the ends of the other species. Thus, we have switched to a system featuring the larger and less secure pen, and the alarm clock set at two-hour intervals, and the muddled brains of the attendants.

If the salmon are running in the deep night in December or January, it is likely that the moon is new, that the river is rising, and that the water is clouded with silt. It is probably raining. The salmon will use these elements of obscurity to hide them from predators while they make a dash toward the spawning grounds.

Tonight it is drizzling lightly, the air full of water just heavy enough to fall to the ground. The drops cut across the beam of my headlamp and seem to be held there motionless, a black-and-white cartoon of rain. In the circle at the end of the beam, the black shag of redwood and the huckleberry understory is everywhere weighted down with water and dripping.

I am in clumsy chestwaders that weigh seven or eight pounds. The rubber bootlegs join at the crotch, and the garment continues up to just above the sternum, where it's held in place by a pair of short suspenders. The suspenders are never adjusted correctly; they are inevitably too tight or too loose. I lurch about like a puppet with too few moveable joints. Longjohns top and bottom against the cold. A Helly-Hansen raincoat and a black knit watch cap put on over the strap that holds my headlamp. To pee, I have to take off the coat, find a place to put it so that it won't get wet on the inside, undo the suspenders and slide the waders down to my knees, unbutton my Levis, and fish around for the fly of the longjohns. The cap can stay on. I turn my back to the river out of courtesy.

The Mattole River runs through the westernmost watershed in California, cutting down through sea bottoms that have only recently, in geological terms, risen up out of the Pacific. It runs everywhere through deep valleys or gorges carved from the soft young sandstone.

Here, only a few miles from its headwaters, the river looks more like a large creek and is closely contained by steep banks. The fish are spooky during this culminating stage of their lives, which is why they run at night, and in murky water. Any light on the water, any boulder clumsily splashed into the stream, will turn a salmon skittering back toward the nearest hole or brushy overhang downstream. She may not try again until another night or, in the worst case, will establish a spawning nest—a redd—downstream from the weir, in a place with too much current to allow her eggs to be effectively fertilized.

I inch down the bank crabwise in wet darkness, the gumboot heels of the waders digging furrows in the mud, the fingers and heels of my hands plowing the soaked wet duff.

On the bank of the river at the bottom of the ravine I hold my breath and let my ears readjust to the sounds of the water. I think I can hear through the cascades of sound a systematic plop, plop, plop, as if pieces of fruit are being dropped into the water. Sometimes this is the sound of a fish searching for the opening upstream; sometimes it is not. I breathe quietly and wait. I continue to hear the sound for a period of time for which I have no measure . . . and then it stops. I wait and wait. I hold my breath but do not hear the sound again. There is a long piece of parachute cord tied to a slipknot that holds open the gate at the mouth of the weir. I yank on the cord and the gate falls closed, its crash muted as the rush of water pushes it the last few inches tight against the body of the weir.

And now that I am no longer trying to sort one sound from another in the sound of the water, it is as if the water has become silent. It is dark. If the world were a movie, this would be cut to black. When I hear the sound I am waiting for, it is unmistakable: the sound of a full-grown salmon leaping wholly out of the water and twisting back into it. My straining senses slow down the sound so that each of its parts can be heard separately. A hiss, barely perceptible, as the fish muscles itself right out of its living medium; a silence like a dozen monks pausing too long between the strophes of a chant as the creature arcs through the dangerous air; a crash as of a basketball going through a plate glass window as he or she returns to the velvet embrace of the water; and then a thousand tiny bells struck once only as the shards of water fall and the surface of the stream regains its viscous integrity.

I flick on my headlamp and the whole backwater pool seems to leap toward me. The silver streak that crosses the enclosure in an instant is a flash of lightning within my skull, one which heals the wound that has separated me from this moment—from any moment. The encounter is so perfectly complex, timeless, and reciprocal that it takes on an objective reality of its own. I am able to walk around it as if it were a block of carved stone. If my feelings could be reduced to a chemical formula, the experience would be a clear solution made up of equal parts of dumb wonder and clean exhilaration, colored through with a sense of abiding dread. I could write a book about it.

The coevolution of humans and salmon on the North Pacific Rim fades into antiquity so completely that it is difficult to imagine a first encounter between the two species. Salmon probably arrived first. Their presence can be understood as one of the necessary preconditions for human settlement. Pacific salmon species became differentiated from their Atlantic ancestors no more than half a million years ago.[2] Such adaptations were a response to their separation from their Atlantic salmon parent stock by land bridges such as the one that has periodically spanned the Bering Strait. By the time the Bering Sea land bridge last emerged, twelve thousand to twenty-five thousand years ago, in the Pleistocene epoch, the six species of Pacific salmon had arrived at their present characteristics and had attained their distribution over the vast areas of the North Pacific. As the ice pack retreated, the species continued to adapt ever more exactly as stocks or races—each finely attuned to one of the new rivers and to recently arrived human predators. If indeed humans first arrived in North America after crossing that land bridge from Asia, the sight of salmon pushing up the rivers of this eastern shore would have served as proof that this place too was livable.

On this mindblown midnight in the Mattole I could be any human at any time during the last few millennia, stunned by the lavish design of nature. The knowledge of the continuous presence of salmon in this river allows me to know myself for a moment as an expression of the continuity of human residence in this valley. Gone for a moment is my uncomfortable identity as part of a recently arrived race of invaders with doubtful title to the land; this encounter is one between species, human and salmonid. Such encounters have been happening as long as anyone can remember: the fish arrive to feed us, and they do so at the same time every year and with an obvious sense of intention. They come at intervals to feed us. They are very beautiful. What if they stopped coming?—which they must if we fail to relearn how to celebrate the true nature of the relationship.

For most of us, the understanding of how it might have been to live in a lavish system of natural provision is dim and may be obscured further by the scholarship that informs us. Our understanding of biology has been formulated during a time of less diversity and abundance in nature; our sense of relationship is replaced by fear of scarcity. By the time the anthropologists Alfred Kroeber[3] and Erna Gunther[4] were collecting their impressions of the life of the Native Americans of the Pacific Northwest, early in the twentieth century, the great salmon runs that had been an integral part of that life had already been systematically reduced. It may be this factor that makes the rituals described in their published papers seem transcendent and remote: ceremonial behavior that had evolved during a long period of dynamic balance has become difficult to understand in the period of swift decline that has followed.

It seems that in this part of the world, salmon have always been experienced by humans very directly as food, and food as relationship: the Yurok word for salmon, *nepu,* means "that which is eaten"; for the Ainu, the indigenous people of Hokkaido Island, the word is *stripe,* meaning "the real thing we eat." Given the abundance and regularity of the provision, one can imagine a relationship perceived as being between the feeder and those fed rather than between hunted and hunter. Villages in earlier times were located on the banks of streams, at the confluence of tributaries, because that is where the food delivers itself. The food swims up the stream each year at much the same time and gives itself, alive and generous.

It is not difficult to capture a salmon for food. My own first memory of salmon is of my father dressed for work as a radio dispatcher, standing on the low check dam across the Sacramento River at Redding and catching a king salmon in his arms, almost accidentally. The great Shasta Dam, which when completed would deny salmon access to the headwaters of the river, was still under construction. Twenty years later, as an urbanized young man, I found myself standing with a pitchfork, barefooted, in an inland tributary of the Klamath River, California's second largest river system. The salmon were beating their way upstream in the shallow water between my legs. Almost blindly, my comrades and I speared four or five of them. When the salmon come up the river, they come as food and they come as gift.

Salmon were also experienced as *connection.* At the time of year when the salmon come back, drawn up the rivers by spring freshets or fall rains, everyone in the old villages must have gained a renewal of their immediate personal knowledge of why the village was located where it was, of how tightly the lives of the people were tied to the lives of the salmon. The nets and drying racks were mended and ready. Everyone had a role to play in the great flood of natural provision that followed. The salmon runs were the largest annual events for the village community. The overarching abundance of salmon—their sheer numbers—is difficult to imagine from our vantage point in the late twentieth century. Nineteenth-century firsthand accounts consistently describe rivers filled from bank to bank with ascending salmon: "You could walk across the rivers on their backs!" In the memory of my neighbor Russell Chambers, an octogenarian, there are stories of horses refusing to cross the Mattole in the fall because the river had for a time become a torrent of squirming, flashing, silvery salmon light.

It is equally difficult to imagine a collective life informed and infused by the exuberant seasonal pulses of surrounding nature over a lifetime, over the lifetime of generations. But for most of the years in tribal memory of this region's original inhabitants, the arrival of salmon punctuated, at least once annually, a flow of provision that included acorn and abalone in the south, clams and berries and smelt in the north, venison and mussels and tender greens everywhere. Humans lived on the northwest coasts of North America for thousands of years in a state of lavish natural provision inseparable from any concept of individual or community life and survival. Human consciousness organized the collective experience as an unbroken field of being: there is no separation between people and the multitudinous expressions of place manifested as food.

But each annual cycle is punctuated also by winter and the hungry time of early spring, and in the memory of each generation there are larger discontinuities of famine and upheaval. Within the memory of anyone's grandmother's grandfather, there is a catastrophe that has broken the cycle of abundance and brought hard times. California has periodic droughts that have lasted as long as a human generation. And there are cycles that have longer swings than can be encompassed by individual human lifetimes. Within any hundred-year period, floods alter the very structure of rivers. Along the Cascadian subduction zone, which stretches from Vancouver Island to Cape Mendocino in California, earthquakes and tidal waves three to five hundred years apart change the very nature of the landscape along its entire length.[5] Whole new terraces rise up out of the sea in one place; the land drops away thirty feet in another. Rivers find new channels, and the salmon become lost for a time.

Even larger cycles include those long fluctuations of temperature in the air and water which every ten or twenty thousand years capture the water of the world in glaciers and the ice caps. Continents are scoured, mountain valleys deepened, coastlines reconfigured, human histories interrupted. These events become myths of a landscape in a state of perpetual creation; they are a part of every winter's storytelling. The stories cast a shadow on the psyche and they carry advice which cannot be ignored. Be attentive. Watch your step. Everything's alive and moving.

On a scale equivalent to that of the changes caused by ice ages and continental drift are the forces set loose by recent European invasions and conquests of North America, the exponential explosion of human population that drives this history, and the aberrant denial of the processes of interdependence which has come to define human behavior during this period.

Somewhere between these conflicting states of wonder—between natural provision erotic in its profligacy and cruel in its sometimes sudden and total withdrawal—lies the origins of the old ways. Somewhere beyond our modern notions of religion and regulation but partaking of both, human engagement with salmon—and the rest of the natural world—has been marked by behavior that is respectful, participatory, and ceremonial. And it is in this way that most of the human species has behaved most of the time it has been on the planet.

King salmon and I are together in the water. The basic bone-felt nature of this encounter never changes, even though I have spent parts of a lifetime seeking the meeting and puzzling over its meaning, trying to find for myself the right place in it. It is a *large* experience, and it has never failed to contain these elements, at once separate and combined: empty-minded awe; an uneasiness about my own active role both as a person and as a creature of my species; and a looming existential dread that sometimes attains the physicality of a lump in the throat, a knot in the abdomen, a constriction around the temples. They seem important, these various elements of response, like basic conditions of existence. I am smack in the middle of the beautiful off-handed description of our field of being that once flew up from my friend David Abram's mouth: that we are many sets of eyes staring out at each other from the same living body. For the instant, there is a part of that living body which is a cold wet darkness containing a pure burst of salmon muscle and intelligence, and containing also a clumsy human pursuing the ghost of a relationship.

I have left the big dip net leaning against the trailer up above the river. I forget that the captured fish is probably confused and will not quickly find its way out of the river pen. I race up the steep bank of the gorge as if everything depends upon my speed. My wader boots, half a size too large, catch on a tree root and I am thrown on my face in the mud. The bank is steep and I hit the ground before my body expects to, and with less force. I am so happy to be unhurt that I giggle absurdly. Why, tonight, am I acting like a hunter? All my training, social and intellectual, as well as my genetic predisposition, moves me to act like a predator rather than a grateful, careful guest at Gaia's table. Why am I acting as if this is an encounter that has a winner and a loser, even though I am perfectly aware that the goal of the encounter is to keep the fish alive?

I retrieve the dip net and return more slowly down the dark bank to the river. Flashing the beam of my headlamp on the water in the enclosure, I can see a shape darker than the dark water. The shape rolls as it turns to flash the pale belly. The fish is large—three or maybe four years old. It seems as long as my leg.

Several lengths of large PVC pipe are strewn along the edge of the river, half in the water and half out. These sections of heavy white or aquamarine tubing, eight, ten, and twelve inches in diameter, have been cut to length to provide temporary holding for a salmon of any of the various sizes that might arrive: the more closely contained the captured creature, the less it will thrash about and do injury to itself. I remove from the largest tube the perforated Plexiglas endplate held in place by large cotter pins.

I wade into the watery pen. Nowhere is the water deeper than my knees; the trap site has been selected for the rare regularity of its bottom and for its gentle gradient. The pen is small enough so that anywhere I stand I dominate half its area. Here, within miles of its headwaters, the river is no more than thirty feet across. The pen encloses half its width. I wade slowly back and forth to get a sense of the fish's speed and strength. This one seems to be a female, recently arrived. When she swims between my feet I can see the gentle swollen curve from gill to tail where her three to five thousand eggs are carried. She explores this new barrier to her upstream migration powerfully and methodically, surging from one side of the enclosure to another. Using the handle of the net to balance myself against the current, I find the edge of the pen farthest from the shore, turn off the headlamp, and stand quietly, listening again.

The rain has stopped. Occasionally I can hear her dorsal fin tear the surface of the water. After a few minutes I point my headlamp downward and flick the switch. Again the surface of the water seems to leap toward me. The fish is irritated or frightened by the light, and each of her exploratory surges moves her farther away from me, closer to the shore.

The great strength of her thrusts pushes her into water that is shallower than the depth of her body, and she flounders. Her tail seeks purchase where there is none and beats the shallow water like a fibrillating heart. The whole weight of the river seems to tear against my legs as I take the few steps toward her. I reach over her with the net so that she lies between me and the mesh hoop. I hold the net stationary and kick at the water near her tail; she twists away from me and into the net. Now I can twist the mouth of the net up toward the air and she is completely encircled by the two-inch mesh. I move her toward deeper water and rest.

There are sparks of light rotating behind my eyes. The struggle in the net translates up my arms like low-voltage electricity. The weight of the fish amplified by the length of the net's handle is too much. I use two hands to grasp the aluminum rim at either side of the mouth of the net, and I rest and breathe. After a bit, I can release one side of the frame and hold the whole net jammed against my leg with one hand. I reach for the PVC tube and position its open mouth where I want it, half submerged and with the opening pointing toward us. I move the net and the fish around to my left side and grasp through the net the narrow part of her body just forward of her tail—the peduncle—where she is still twice the thickness of my wrist.

I have only enough strength to turn the fish in one direction or another; were I to try and lift her out of the water against her powerful lateral thrashing, I would surely drop her. The fish is all one long muscle from head to tail, and that muscle is longer, and stronger, than any muscle I can bring to bear. I direct her head toward the tube, and enclose tube and fish within the net. I drop the handle of the net, and move the fish forward, toward the tube.

There is a moment while I am holding the salmon and mesh entwined in elbow-deep water when everything goes still. Her eyes are utterly devoid of expression. Her gills pump and relax, pump and relax, measured and calmly regular. There is in that reflex an essence of aquatic creaturehood, a reality to itself entire. And there is a sense of great peacefulness, as when watching the rise and fall of a sleeping lover's chest. When I loosen my grasp, she swims out of the net and into the small enclosure.

Quickly, trembling, I lift the tail end of the tube so that her head is facing down into the river. I slide the Plexiglas endplate into place and fasten it, and she lies quietly, the tube just submerged and tethered to a stout willow. I sit down beside the dark and noisy river, beside the captured female salmon. I am sweating inside my rubber gear. The rain has begun again. I think about the new year and the promise of the eggs inside her. I am surrounded by ghosts that rise off the river like scant fog.

### Notes

1. This temptation seems to be a regional institution. See Alfred Kroeber and Samuel Barrett, "Fishing Among the Indians of Northwest California," *University of California Anthropological Records* 21:1 (1960), an encyclopedic approach to indigenous fishing technology, which indicates a more or less even split between using weirs to direct salmon to a fishing platform (only functional when manned by a spear or net fisher) or to a pen checked periodically.

2. See Ferris Neave (Canadian Fisheries Research Board, 1958), quoted in Anthony Netboy's *Salmon, the World's Most Harassed Fish* (London: A. Deutsch, 1980).

3. Kroeber's work in physical and cultural anthropology among indigenous people of California is legend. See his monumental *Handbook of the Indians of California,* Bulletin 78 of the Bureau of American Ethnology of the Smithsonian Institution (Washington, D.C.: U.S. Government Printing Office, 1925). My copy is a reprint by Dover Publications, 1976.

4. See the work of Erna Gunther, the great ethnobotanist of the Pacific Northwest, especially "A Further Analysis of the First Salmon Ceremony," *University of Washington Publications in Anthropology* 2:5 (Seattle: University of Washington Press, 1928).

5. Researcher Deborah Carver has related Yurok "myths" of land formation and the destruction and dispersal of peoples as evidence of a massive earthquake on the Cascadian subduction zone within tribal memory. More recent studies have dated the last massive movement of the Cascadian subduction zone as occurring on January 26, 1700.

## Everglades

*Ricardo Pau-Llosa*

—For Barbara Neijna, after her sculpture/installation, *Under the Green Coat Is an Endangered Heart*

We have always been ready
for the soft killing
signed by boot tread and post.
The songs were of freedom and money then,
and they kept us clean as we cleaved
the earth and drank it coin dry.
We never caught ourselves
in our concrete webs, never saw
the eye of conscience crowd out
the anger of traffic
in the rearview mirror.
All around us the sprinklers
threw rainbows around like bad checks.
Their music twittered, byzantine.

We never caught on
atop the incandescent skyscrapers
that the horizon was dying.
The thought that we kill by naming
is criminally fresh.
Your metal cone points downward,
whispers its spiral,
sky-heavy
muse.
And we, the alp-less, insist
on scaling the ever resting clouds.
We keep looking up
as if the sky beckoned the city
to crane its neck more and more,
to leave its helipads like eggs
on the infinite vapor.

We never caught
the need in time, and yet you point
down, beg us to look down,
your cypress saying enough
of sky. Enough profit
and order. It is the earth
we walk upon and cannot love,
but it is there
that our mirrors trickle.

Kristin Ordahl, *Water Glass Series,* 2000, video still.

# Waters under the Earth

*Arthur Versluis*

How rarely we consider what flows under the surface, the mysterious currents of water deep below us in the earth. For many of us, the landscape above is all we notice; we are captivated by what we can see, and pay little attention to the waters that run deep in the ground. We observe the rain or snow as it's coming down, but rarely do we ponder where the water goes once it has seeped into the earth. With a prosaic term like *groundwater,* we act as if we can analyze and make sense of what is happening far below us, when we still do not fully understand water's mysteries. And from what I've seen, it seems that most of us are not interested in knowing. In our lives are ciphers that we may not want to solve.

So it was with me, until one day a few years ago, when I saw a realtor's sign along a rural road in my native mid-Michigan and followed its arrow to a gravel road lined with wide-trunked maple trees whose foliage formed a canopy under which I drove. On the left were a white-sided farmhouse, verdant alfalfa pastures, sandy arenas, and two barns, both newly built. On the right was a huge red older barn, and below it, into the distance, rich alfalfa on rolling hills along the river valley. Sleek thoroughbred horses looked up from the pasture next to which several children were playing. Never would I have suspected, as I walked toward the genial-looking fellow in his late thirties standing near the barn, that beneath our feet flowed poison, silent and inexorable.

But there was something incongruous in this seemingly idyllic setting. The fellow's cars and truck were parked in front of the older barn, not in front of the house, and his children behaved as if this barn was their home. And, it turned out, they indeed were living in the barn then. The house, which he and his wife had bought and totally restored, was now for sale by the bank. They had built two new barns across the street, one for raising their horses. Over the ten years they lived there, they had had six children. Dan and his wife, Sara, had bought several hundred acres of this land and had been living their dream. Though they both worked at outside jobs (like most struggling farmers in America), they had managed to build a good working farm. And they had had eighteen horses the past spring when the first government technician, wearing gloves and outfitted in what looked like an astronaut's suit, showed up without warning and announced that he was there to test their groundwater.

The results were back within a few days, yet the government would tell Dan nothing. Somehow only the bank was privy to just how poisoned his well was, and they weren't telling. In the meantime, Dan called around to the state Department of Natural Resources (DNR) and to other farmers, trying to find out what was going on. The story that he pieced together over the next month stunned him: A landfill that had been a dumping ground for toxic industrial wastes since the 1960s was located less than half a mile from his farm.

At first, no one acknowledged that anything much had gone on there. The local DNR officials told him that there was nothing to be concerned about. Only when he finally called Washington, D.C., did he learn, to his alarm, that the dump up the road, once called "B. and B. Landfill," was on the so-called Superfund list of federal emergency sites, labeled "extreme urgency."

That site was forty acres of rolling land dotted with artesian springs just south of what would later become Dan's and Sara's farm. According to Dan, in those days, the 1960s, the local government figured that since the land was unsuitable for farming, it only made sense to use it for dumping. The dump was listed officially as taking only household trash, tin cans, and old bottles. But in truth, like similarly labeled dumps throughout rural America, it had been a surreptitious toxic waste dump. And although it was listed as being clay lined, it actually had no lining at all.

It gets worse. Much worse. As Dan searched the DNR and Environmental Protection Agency (EPA) records—tens of thousands of pages—he discovered that men had admitted, in sworn testimony, to having driven tankers carrying sixty thousand gallons of heavy metal sludge onto that land and just opened the spigots.

"What kind of men would do that?" Dan asked me with a long, searching gaze. "I mean, what kind of man would do that? They knew what they were doing, even back then. And the local officials knew too. Some of them were paid off. This thing goes a lot higher. There're some powerful people involved, not just bribed city peons like mayors, but factory owners and the state Department of Corrections, which dumped plenty over here at night and on weekends. Do you think," he asked, "that the state Department of Natural Resources is going to prosecute the Department of Corrections? I don't think so. And those powerful people have managed to get some of the records sealed so you can't find out who they are. I talked to a DNR hydrologist who had seen some of the sealed records," he added. "She said, furtively, that it was far, far worse even than what I had seen in the documents I saw."

All of this testing and testimony had been going on during much of the time Dan and Sara were building up their farmstead nearby. For at least six years, the federal EPA officials, the state DNR officials, and the frightened city officials all had been driving back and forth past Dan's land, had been testing the soil, had put the site on the federal emergency list, and had taken thousands of pages of testimony. Never, not once, did any of this leak into the local newspaper, nor did any one of those officials ever tell Dan that he ought to check his water. Not one word. In fact, even when he did find out and he called the DNR and the EPA, he was met by barriers of abstract jargon, phrases like "health-based risk criteria ratings," designed to insulate people from what was really going on. No one spoke in plain language; no one warned him.

During this time, Dan began researching at the local DNR office, and read through countless pages of testimony. There, in sworn affadavits, truckers and factory owners were laughing at the investigating federal officers—laughing, right there in the account. At one point, a frustrated federal official said to the man being questioned that he had a choice: he could have dinner that night at home or in a fast food joint on his way to federal prison. That woke the recalcitrant fellow up some. This alone was deeply disturbing, but what upset Dan most was the fact that the local and state DNR officials and the federal agents had known about this for six years and never said a word to him or to any of his neighbors.

So he called up the director of the DNR.

"Do you remember this site?" he asked.

"Sure," was the reply.

"And all those times you drove to and from it, do you remember a white farmhouse and big red barn on that tree-lined dirt road?"

"Yes. Is that your place? Quite beautiful."

"And did you see those six kids playing in the yard?"

"Yes."

"Do you have children?"

"They're grown, but—"

"But you have children of your own. Now imagine this was your place and those your children. How would you feel? But you never thought to stop. You never told us that we might die from that toxic waste a half mile away."

"No. That's not my responsibility. It's not my job; it's not in our mandate."

"Really. And who are you serving? Last time I looked, we were paying taxes."

"I'm sorry, we only collect information. If you were to call, we'd tell you."

"Right. 'Honey, I think I'll call the DNR today and see what's on their minds. See if somebody's put us on a federal emergency list or something. Heck, you never know.'"

Surprisingly, the entire affair had begun more or less by accident. Six years prior, some people down near the river had begun to get sick, and when the water was tested, health officials found that it was laden with toxic chemicals from a leaking dump nearby—just down the hill from Dan and Sara's farm. Federal EPA officials were called in, and a host of local and state officials had gathered at the site near town to greet them, some of whom bore responsibility for this fiasco in various ways. The dignitaries waited and waited at the dump near the river, until someone figured out that the EPA people were up at another crossroads, testing a different location. Everyone drove over to tell the agents that they had taken a wrong turn, only to find that their error had uncovered toxic pollution even more dangerous than what was at the site nearer the river.

But news of these events never reached Dan's and Sara's ears. It wasn't until the spring of 1995 that they found out, when they were intending to refinance to purchase another seventy-five acres, and the banker, looking on a map, informed them that it appeared their land was on a federal emergency site. "*What?*" Dan cried out. The banker's inquiry at that time sparked the EPA groundwater testing, the results of which then were withheld from Dan and Sara. Accidental discoveries seem to have played a large role in all of this—

along with evasion, concealment, and the refusal to reveal the truth. Everyone involved seemed bent on withholding what they knew, most of all from those directly affected.

While Dan and Sara were slowly discovering the extent of the danger and the conspiracies, they were also talking to friends and acquaintances. They found that their situation was not unique. An attorney in his early sixties who had owned a horse farm in southern Michigan worth nearly $2 million, which he had built up over years and years, was preparing for retirement and a quiet life of raising horses when he began to suspect his water because his mares gave birth to stillborn and deformed foals. He sent water samples to the local health department for testing, not knowing that the health department would notify the federal government if certain chemicals were found. One day, without any warning, an EPA official appeared at his farmhouse and told him that he had twenty-four hours to vacate the property—permanently. He lost everything. His advice, after years of litigation? Just write it off. Walk away from it, or you'll never have peace of mind again.

Dan and Sara might have ignored this advice were it not for a couple they met from Kentucky, both physicians, who raised horses in a rural area. In spite of their medical background, they couldn't figure out what was causing the spate of abortions among their mares, nor did they know why, the following year, all the foals born on their farm were either stillborn or deformed. Certainly it wasn't their bloodlines that brought about the six-legged thoroughbred colt. The following year, their children began to suffer from what is euphemistically called "failure to thrive." They had them tested for all kinds of ailments, but nothing showed up. Yet the children grew more and more ill. It wasn't until a friend suggested that they have their water tested that they discovered the truth, and by this time, their children had developed permanent liver and kidney damage. They too were forced off their land and had to abandon it.

But the story extended far deeper. Underneath Dan and Sara's land, hidden currents of poison were leaching through the ground, spreading down toward the river valley below.

"There is nothing we can do," he said, "but watch it flow downhill. The hydrologist told us that conditions here are ideal for catastrophe. The toxic waste was poured by the hundreds of thousands of gallons directly into land dotted with springs, signifying the entry point into the groundwater that in turn is drawn by gravity into a basin under the rolling hills sloping down to the river valley. Along the way, every single house will be affected: next our neighbors, then those people across the street, finally people by the river down below."

Curiously, when Dan went from house to house to tell his neighbors the truth, no one wanted to listen. A couple of them actually shut their doors on him.

"Everyone," he said, "wants to hide under a rock. They said that they'd wait and see, or that nothing would happen. But you know all too well," Dan added, "that late at night, when that couple is lying in bed, they talk about it, and you know that deep down they must recognize what's going to happen. In ten or twenty years, all these houses will be empty and abandoned. The bank can never sell our place, and they can't sell theirs. People aren't going to live here without showers or baths, without drinking or cooking water, and they're all going to have to leave. I expect I'll come back here at some point and see the roof of our house caved in. It's sad, and no one wants to face the truth."

Then our conversation adopted a broader perspective as we considered the larger implications of this situation.

"I believe in the free market as much as any other American," said Dan, leaning against the white fence rails and watching his daughter climb a tree. "But the people who did this, they're sick. You know what it is?" He glanced toward me. "It's the free market without religion. No morality, just money. That's what evil is, here in America.

"It's what gave us all these things we have. But you know, it's the same thing that can destroy us. We've got to learn to live in a different way. And I think it's possible. It's got to be possible. In the meantime, it's true we've lost our farm, but we've still got our health and each other. We can start over. Maybe in the end we'll see this as good, even an opportunity."

The truth is, though, that poison runs beneath the surface of all our lives. Like Dan's and Sara's neighbors, most of us don't want to acknowledge, much less face it; we'd like to think that what we don't see doesn't exist.

Consumerism casts a long shadow. All that plastic, all those industrial chemicals, the waste from the prison industry, medical waste, heavy metal sludge, pesticides and herbicides, all that smoke and poison has to go somewhere. And since we would prefer that it go somewhere out of sight, it generally does—down beneath the surface of the earth. If it was once true, as Emerson said, that each of us thinks of a farmer as farming for him by proxy, so is it true today that each of us has had a toxic waste hauler dumping surreptitiously into the groundwater for us by proxy. We may not do this ourselves, but we willingly participate in a system that fosters such destruction. Every one of us bears indirect responsibility for the poisoning of Dan's well.

That said, some do bear direct responsibility: city officials who took bribes or accepted favors; county and state officials who looked the other way; and most of all the factory owners, the prison officials, the contract waste haulers, and the men who turned the spigots on the tankers. They knew what they were doing. Catastrophes always begin with a devil-may-care attitude and actions that at the time seem not to bear serious consequences. It is only over time that we come to realize the gravity and extent of the repercussions we have incurred.

A story like this may seem hard to believe, and to this day many in the vicinity of Dan and Sara's farmhouse still refuse to believe it. I stopped by their former farmhouse on a recent sunny Sunday afternoon to find out what had happened. To my surprise, there were children playing in the yard, two of them leading around pale Shetland ponies. I went to the front door, where I met Mary, a friendly woman in her forties. I introduced myself, and we talked for a time about what it had been like to move into that house.

The farmhouse had stood empty for several years since I had first met Dan and Sara, when they and the children were living in the barn across the street. But eventually Mary and her husband, Jack, had purchased it from the bank. I asked Mary if they had had the water tested. She responded that they had had the usual health department tests. But since the health department doesn't test for the kinds of groundwater pollution that come from toxic industrial waste, I wondered if Mary and her family were concerned.

"No," she said, "we never really think about it. We've got a filter on the water. And anyway, the neighbors think that the fellow who lived here before was a little daft. Besides, the bank wouldn't have sold us this place if the water wasn't safe, would they?"

The bank's priority undoubtedly was to sell the property, I told her. She replied that she and her husband just were not worried. And I wondered to myself: If someone wanted to purchase a house just down the hill from a former landfill and federally recognized toxic pollution site, then who was to stop the purchase? The state had no interest in raising public alarm, obviously, since the pollution at the site came in part from state correctional facilities nearby. Local officials certainly wouldn't be interested in stopping the purchase, since it would eliminate some of the local tax base. Further, the neighbors themselves did not want to hear about having their water tested, nor did they want to investigate what happened at the landfill just up the road. They simply wanted to go on with their lives, as this couple did.

And so we are left to wonder what the truth is. Dan and Sara now live in the next town over, poorer for the loss of their farmstead, but happy that they no longer have to be concerned about their water. Jack and Mary live in the farmhouse that Dan and his family had abandoned, and all seems to be well. The land looks healthy on the surface—those green rolling hills of flourishing alfalfa, corn, and soy.

But what is happening below? That there was an illegal landfill just to the south where, thirty years before, illegal toxic waste dumping had taken place, that much is certain. To this day, the land is listed on a federal registry. Yet what happened to those tanker-loads of poison that were poured out on the swamp? Are we truly to believe that all is safe? That beneath the earth, the poison has simply vanished, run off?—into what? where? Do the waters under the earth run brackish and foul, like poisoned blood? I have no answers for these questions. No one really knows what is in the waters down below, or where they run, whether poison is slowly and quietly seeping down toward the river a mile or so away.

Perhaps Jack and Mary and their neighbors are right. Perhaps it's better not to know, to live blithely unaware. But one has to wonder what kind of legacy we have left for ourselves and our children. Such were the thoughts I had as I left that farmhouse. But before long I was back at home, once again immersed in the details of daily life—the bills to pay and the tasks to finish—so busy that when I filled a glass of water from the kitchen faucet, I hardly gave it a second thought.

## The Bronx River

*Bob Braine*

Luminescent plastic bags float suspended under water like unwitting Seechi disks. I think about a sea turtle eating one in the open ocean, mistaking it for a jellyfish, and choking—like a child playing with a dry cleaning bag—spiraling down to the bottom, another victim of a common household accident.

As a predator I have the same relationship to the plastic bags as the sea turtle has. I am looking for life, and the sight of one of these pale bags pulsating in the current triggers those instincts—squeezes my adrenal glands tight and changes my blood chemistry.

Like the turtle, I want this plastic bag to be alive—somehow it is the missing aquatic mega fauna link that the Bronx River needs. Big life.

Traveling on the water in an urban environment, the sounds one hears are certainly different from those of the jungle—gone are the harsh cries of macaws, the splash of big fish, the meow-like barks of the toucan. What you get is the dull hissing roar of traffic moving on bridges overhead. Instead of downed trees blocking the river, you are faced with shopping carts lying on their sides in the muddy water, functioning like giant strainers and somehow reminding me of an animal that has gone down to the river to leave its metal skeleton in the brown water. A final pilgrimage. Parallel evolution.

# The Lessons of the Well

*Linda Tatelbaum*

In a drought, there's only one thing to do—wait. We tried collecting rain in buckets. But there was no rain. A line of white plastic pails sat hopefully under the roof line, the heat slowly turning them green on the inside. The garden grew, somehow, except for the lettuce, planted and replanted, but it just won't germinate in dust. Then the well went dry. Dirty dishwater, saved in a watering can, only goes so far in a big garden, and there is less and less waste water as we tighten our usage more and more. Hauling water from town guarantees that.

And so I'm down here, eye-to-eye with the rocks in this old well that should be filled with beautiful water but isn't. I'm nine feet down, inside the circular stone cylinder the old-timers laid up two centuries ago, my feet on the smooth clay bottom where, until an hour ago, the muck was thick. Kal and I took turns climbing down the ladder and handing bucketloads of debris up. We're hoping the veins that feed the well are clogged, not really dry at all. So I'm down here, with no ladder now, standing on the bottom of the bottom, in the place that waits.

I'm here because I just can't wait anymore. I'm too worried. Everything is drying up in the drought, even my creativity, and I know you can't push the river, but you can try, can't you? I still go to the desk every day, as if writing something, and even though nothing comes flowing in, I've got to keep those buckets lined up just in case. A drought can end any time, without warning.

This spell without rain, record-breaking, heart-breaking, leads me to wonder if it ever will rain again, or if this is it—the turning-point in planetary viability, the end of lettuce as we know it.

I've got an old pail down here, one found thrown in the woods from long ago, maybe for hauling water, maybe for milking. No tool will do what I have to do today, no tool but hands, fingers, arms. I hesitate, then squat down to the lowest circle of rocks, the biggest ones, and lord knows how they got them down here and rolled them into place so perfectly, all done in a hurry before water started to flow back in. I don't have that problem. I have only the fear of what can't be seen hiding in dark crevices between the rocks, blocking veins that might, just might, carry water into the cylinder if we clean them out.

What do I fear in the accumulated mud of history? Something dead. Or, worse, something alive in the moss, algae, pine needles. Spiders and their eggs, frogs alive or dead. Snakes? The skeleton of some drowned rodent. Glass.

I reach in, gingerly, with the edge of one palm, and sweep a slimy load into the pail. No, it isn't slimy. It is silky, silty. It is lovely and smooth and cool. I reach in again, deeper this time, and let it ooze through my fingers as I drag out more. And more, up to the elbow by now, and wanting to go deeper, deeper, wanting to grab those veins and milk them into the waiting pail.

I am inside the round space these stones make. But watching my arm disappear into the dark, into the inside of the inside, I see that I am outside, really, separated from earth by these rings of stone. I look up and see Kal kneeling down at the rim, a bearded face topped with an old green cap spattered with paint. To him, looking down at me enclosed in this narrow hole, I must seem inside. But he's the one who is. He is inside the usual space of sun and sky that I am out of, that I am under.

"Ready?" he asks. His faraway voice bounces off the deep, dry stones. He stands up. His face disappears, and I watch legs grapple for balance on what, to him, is the rooty ground, a roof top to me. He tugs on the empty hook at the end of a perpendicular pole hanging from the long boom of the well sweep. The boom pivots in a crotched oak, and the pole comes down toward me into the well. He leans his weight on it, counterbalancing the rock lashed to the far end that weighs more than a full bucket. I hook the bucket of muck to the pole.

"Okay!" I signal, and press my back against the cool stones. He lets up, and the pole rises effortlessly by gravity as the counterweight slowly drops to the ground and lifts the heavy bucket out of the well. He goes to empty it in the woods. I wait, pondering the contradiction of gravity lifting something. His returning footsteps thud above me, and he clips the bucket onto the hook and yanks the pole to within my reach again.

"Pretty slick!" he shouts down.

"What, the muck?" I yell up.

"No, the system—this well sweep. You were right!"

"The muck's pretty slick, too. My hands feel great. Maybe we should market this stuff!"

Wait. Did he say I was right? Don't tell me, after twenty years of marriage, we're finally learning to work together? When we first settled on this wayback land, with no skills, we used to fight about how to do things. Figuring out a job meant power, and in those early days it was power-up, power-down, the winner gets to be the leader, the other the go-fer. Now our style together looks more like this clever use of gravity, lifting weight by dropping weight.

I said we should rebuild our old well sweep, which had rotted and broken, before cleaning out the well. He said it was too much trouble, we should just use a ladder. He was probably remembering my tantrum years ago, building the first one. My crisis moment, that morning down in the mossy vale where we'd found this old well. No water, no electricity, no house—the consequences of our choice to live in the woods came crashing in on me. *Fool!* I yelled at myself. How am I supposed to figure out how to get this thing balanced? I've studied history, poetry, Latin, French. I flunked physics. Never even heard of a well sweep till now. How can I test this rig without actually building it, and how can I build it without knowing how high to make the fulcrum, how long the boom, how much damn counterweight a bucket of water needs? *Idiot!* I have an advanced degree in *nothing!*

I came to the right place to discover that, anyway. And to earn the *n*th degree in all things practical. Including marriage. I thought his idea of using a ladder was stupid. Why take up half the hole by sticking a ladder down there, and lugging every bucket up the slippery rungs, when physics can do it for you? But instead of saying so, as I would have before earning my *n*th degree, I just went ahead when he wasn't home and rebuilt the well sweep, very easily this time, and surprised him with it today, walking down here to start the job, carrying a ladder between us. He admired it, then proceeded just the same to lower the ladder into the well.

"Well?" he said. "Me first?" And down he went with bucket and shovel. I rolled my eyes at the disappearing top of his cap. He stood on the lowest rung and leaned down for the first bucket of muck. I figured it would only take a round or two of climbing up and down before he saw things my way.

"Try standing on the bottom," I called down, seeing his shoulders twist painfully as he stooped for each shovelful. He tried stepping off the rung and, with a yelp, went up over his knee in muck.

The ladder, it turns out, was the perfect tool for the first stage of the job, a solid platform to stand on. And now that we've arrived at the hard clay bottom, the well sweep is just the thing for the rest of it. Down, up. Pull, push. Wrong, right. Forces work together to get things done, in physics as in marriage.

In writing, too, which is another way to get things done, inside and out working together. It's just that sometimes you're at a loss, can't find the leverage, no force to begin. If you call patience a force, that's the one I exert every day during this creative drought. Patience is readiness in a state of tension that looks like rest. Rest while keeping an alert mind, rest with one eye open, rest while listening for the first pattering drops in an empty bucket. Patience exerted over time produces another force—desire. As I reach into the yielding muck, is it water I thirst after, or a way to understand the bottom?

Deep in this most unlikely place, in the place with no water, in the place known rock for rock only by the old-timers who touched them all, I'm looking for the inside of my outside, what matters to me even in times of drought. When the well is full, I forget the work these rocks do to hold water in one spot for use. I stay at the house, and let our solar-powered pump run the tank full. Back when I carried water every day using the well sweep, an oaken bucket, jugs, and a yoke, it was hard to forget the eternal cycle that carried me along, now lugger, now drinker, now lugger again.

From where I stand contained inside this empty cylinder, chilled by the breath of absent water, it's easy to recall the value of what isn't. The loop is broken. Water has to come from somewhere, or you die. Transporting it here in gallon jugs from faucets in town is an ordeal of lifting and carrying it home, then lifting for each use, right down to the smallest sip of water, or rinsing a toothbrush, or washing the canning jars, or watering the animals. Arms, arms, arms, in this harvest season already asking for more than arms can give. Is it time to forgo the values we've learned as homesteaders on this land—to cope inventively, spend frugally, honor the past—and have a well drilled? An honest-to-god hundreds-of-feet-deep well would give gallons per minute and never run dry. We weigh this question as we climb up and down the ladder, as we fill and empty the pivoting bucket, as we clean the muck off the hard bottom.

The bottom is the source of patience, and patience is a force that, over time, produces the story of a life. I've cared for this old well, and it has nurtured me. Will I now leave it all alone down here, where the cedar cover will rot and fall in, broken branches will cover the circle of stone, rocks cave in, animals drown? I've seen what happens to abandoned wells. I've had to climb down into a few of them and clean out the broken glass, rotted wood, rusted metal. I've immersed myself in the old settlers' hopelessness as they gave up and pitched it all in the hole and drove off in their wagons for an easier life elsewhere. Shall I, too, give up on this old well to get an easier life? Kal and I come up from the bottom with the answer, which is also a question: If you're not grounded in your history, then what are you?

I'm traveling the ribbon of this story as it curves in a never-ending loop. Is it about the water, the drought? Is it about marriage? Is it the tale of a writer at the bottom of her dry well? No matter which side of the Moebius strip I glide along, I always come out on the other, with no place to say, Ah yes, now I'm talking about what it is, now about what it means. Now an empty well, not what it contains. Now a jug of water, from which I drink.

$V = \pi r^2 h$. It all comes down to a formula. Say what you want about history, but can you drink it? $V$ is for volume, the volume of a cylinder, without which you die.

Another formula could have the water gushing in a matter of hours. Just add money to a diesel-powered auger, and we might end up with five gallons a minute, more water than we'll ever need. Instead, we decide to hold onto our money this time. We'll wait. Patience again, a force to reckon with. I clamber out of the clean well one last time. We kneel by the rim and peer at our faces reflected in the 34-inch circle of gleaming clay, 3.14 x 172 = 907.46 square inches of no water. A flat circle is not enough. It needs volume, like the accumulated pages of writing that become a book. It needs depth. Height. We wait.

One hour later, half-an-inch, or two gallons. Enough to do a small batch of dishes, have a cup of tea, wash our faces. In other words, not enough. Nineteen hours later, we have 10 inches or 40 gallons, four days' worth. We start using it, cautiously. It doesn't recover all the way. We go back to hauling from town for a few more weeks. Then the autumn rains come, and the veins open up. We have eight feet of water in the well.

*V* is for volume, without which you die. But solid geometry doesn't tell as much as history after all. It doesn't give depth to this cylinder created by the dead who, in life, worked together to lay up these stones one by one. How much does this well hold? The image holds everything I want from it, a whole volume. I let the words trickle in, chapter by chapter, the first of which is always rock.

Rocks got here first, by a force one can scarcely imagine. Rocks come before the words to tell their history. Words are never quite it anyway. A word starts somewhere, and the rest escapes, leaving a residue of desire. What we want drives our history, but who can say what rocks wanted, in landing here? Patience is time metered by human desire—waiting—waiting *for*— This clock can't measure the patience of ice sheets heaving and scraping to carve valleys studded with what they drop. Slowly, ever so slowly, without waiting for, without wanting, without a plan, they thaw and move toward what oceans do not yet exist. Solid becomes fluid, inside becomes outside, the core becomes the surface upon which we will build, and the materials.

But first, before building, you have to go back under, to what really matters. A man, two hundred years ago, stumbles over scattered rocks as he paces with a Y-shaped dowsing rod. He can't sustain a life on top of these acres until he goes down below them and finds water. Tipped off by a puddle in the wet season, or a growth of ferns in this one spot, or maybe a willow shoot, he follows his rod back and forth. It might have been a dream he had, of cows clustered in this shade, luxuriating in a long drink on a hot day. It might have been a little girl who whispered in his one good ear and pointed a chubby finger at the place. The tip of his Y bends suddenly to the earth, and "Here 'tis," he declares, just where the little girl had said she smelled it.

And so the men take up their shovels and begin by chopping open a layer of sod, digging till rocks impede, prying them up with picks and bars, and piling them to the side. With flat-bottomed sleds they drag the removed dirt to the ledgy garden site where potatoes should grow, but won't, without depth of soil. If time is a force that produces layers, each occluding the last, it takes some other force to peel one back and lay it on anew. The force of thirst and hunger, the desire for water, for potatoes. The imperative: *Live.*

They dig for a week or more, a crew of men, all brothers and sons and cousins. Dig until dark, dig against time, for with time the water will fill the hole where they need to stand, each man taking his turn. Down in the wide hole, five feet in diameter, he bends to fill the pail, then stretches back and watches the bottom of each bucket as his brother hauls it up with a block-and-tackle.

When, finally, the firm layer is reached, "Ho!" he calls up, and they lower the biggest of the rocks, one at a time, to form a circle on that deepest clay. Each rock, handled once to remove it from the earth, once again wrestled back in place on the hard bottom. Each rock known by edge and girth and texture. Known by ache of back. One perhaps known by the wicked language it draws when its mass plus gravity collide with a man's finger.

Up and up they lay the interlocking circles, till soon the man inside can stand on stone and keep his boots out of water. A week of wet feet, wet socks steaming on a rack above the cook fire, is all any of them can stand. The women, too, waiting for their water, making do with collected rain and runoff scraped by half-buckets from the shallow ditch by the road. The cylinder reaches the rim of light, the last man scrambles out, and they finish the edge with large flat stones. By winter, the water will hide their work. By winter, the man will have learned to live without that finger, crushed in the effort. Worth it, though. Can't live without water. Can live without a finger.

The first chapter in the human story, rock, demands a compromise between vision and material. Build what you *can*. Wood is chapter two. Wood can be shaped to human measure with sharp tools. Trees can be felled, bark stripped. Build what you *will* with your level lines, flat planes, square angles. Build as straight and upright as an I.

A wood cover, then, for the well: square frame of cedar, boards, hinges to open the lid. Drop a wooden bucket down at the end of a wooden pole that hangs from a wooden boom that pivots in a wooden crotch. Wood carries wood down into water contained by stone. Wood grabs its portion of water and brings it up.

Wood's chapter continues the story of life on the land. It encloses families under roofs, behind doors. It lays flat for their tables and beds. Curved, it lets them rock in chairs by the fire of its burning. Hollowed out, it serves their food. It gives them fences to restrain the cattle, barnboards to house the hay and coop the chickens. Shade, firewood, tool handles, buckets, barrels, baskets, sleds and snowshoes, wagons, stools, toys. And none of it lasts. These homesteads, a whole road of them built by the settlers of 1777, burned down one at a time with all they contained, leaving only the rusting metal blades and cups and stove lids, only the stone walls, the crumbled brick chimneys, the gravestones. The charred remains of chapter two, left to rot. Rot is chapter three, the longest of all, with no one to tell it.

I will tell it, then.

They were Hessians, the first ones to settle this ridge so far from their "Vaterland," Germany. Upon the orders of King Friedrich II, the young men had been taken from their fields while mowing, shirtless, to be sold into the service of King George III, who needed men to help the British fight the American colonists. Their fathers were given $35 for each lost son. The sons were given uniforms, green coats and black hats trimmed with gold braid, and white pants. Shipped to America. The pants didn't stay white for long in battle, and the British surrendered, and the soldiers were taken prisoner in Boston. Paranoid, they sailed Down East to the German settlement in Waldoboro, Maine. From there they paddled up the grassy Medomak River, no longer belonging to the English. American now, they found land of their own on this ridge where they would not be snatched from their haying to never come home for dinner.

They cleared fields. Cut lumber. Built houses and barns. Planted garden rhubarb, asparagus, apple and plum trees. Raised new generations, one every eighteen years or so. Built a one-room schoolhouse. Opened flour mills and barrel mills along the Pettingill Stream, the Medomak River, the St. George. Blacksmith shop, dry goods store, church. In 1848, the children died of smallpox. The elders got older with no one to take over. Some stayed to age and die on the farm, buried in the cemetery next to their babies who never produced grandchildren. Some moved off the ridge, and spread their names to the neighboring towns. Sukeforth was one, whose ancestors dug our well. What could he do? The house burned down. For all I know, the well went dry on him, too. They left no words to record their history. By the 1870s they were all gone, and only the stone cellarholes remained, the lilac bushes, the rocked-up springs eternal.

**Warm Rain**

*George Keithley*

Adam David Clayman

Night and day the rain spills its silence
over us. South of Galilee not one hand
tends the orchards; the pipe-fed fields
idle on the floor of the Ghor
or *Al Ghawr* absorb the dripping stillness.
Twelve days' rain drenches the valley.
High water in the *wadis*. Only
these girls wrapped in gray wool
each morning gather by the river.

Brooding sky. Gravel soil
swept away, water
roiling downstream,
splashing into the salt
slow sea. Green reeds
glowing. Two girls lean over

the bank. Laughing, they lower
their bare feet into the flow.
Together they unbundle
rolls of soiled clothes
to catch the current
flapping past. Upriver
the sky flashes silver—
A stand of slender young
willows washing
their hair in the high wind.

Under the rain the river
Jordan rises, trembling,
the lean brown arm
of the dead
god growing restless in the reeds.

## Morton Salt Disaster

*Ted Steinberg*

*Question:* Can a large dose of Morton Salt cause a flash flood to occur? A flood, moreover, that swept 238 people to their deaths? It is by all accounts a very odd question to be asking. But answer it we must to fully understand the strange places that our will to dominate nature can take us.

The flood in question struck Rapid City, South Dakota, in 1972. At the eastern edge of the Black Hills in Mount Rushmore country, Rapid City takes its name from a white-water stream (Rapid Creek) that plunges from the hills right through the town. In the late nineteenth century, settlers shied away from building homes too close to the stream out of fear of flooding. But in the years after World War II, as the population of the city swelled, private homes and trailer parks squeezed their way onto floodplain. Such development was given a big boost by the Bureau of Reclamation in 1956 when it built Pactola Reservoir for controlling floods. In 1972, some 43,000 people had created what must have seemed like a comfortable and safe existence beneath the bureau's dam and water control project.

Early in the week of June 5, 1972, a tropical air mass pushed its way across the western United States. Flash flooding occurred in the Mojave Desert, burying cars under piles of mud. Heavy rains battered Bakersfield, California, which experienced significant flooding and damage. This powerful and huge air mass then proceeded north and east until it arrived in the Black Hills, precisely in time to clash with a stream of very cold air flowing down from the Arctic. The collision of the two air masses on June 9 produced a series of thunderstorms that just would not quit. Some areas in the Black Hills region received fifteen inches of rain over the course of less than six hours. Fifteen inches. According to one government report, the rainfall west of Rapid City "averaged about four times the six-hour amounts that are to be expected once every 100 years in that area." That is a lot of rain. Even worse, much of that rain fell on the roughly fifty square miles of land area *below* the Pactola Reservoir. Thus the bureau's flood control structure was about as useless as a fold-up umbrella. A flash flood started, and when it was over 238 people had died. The bodies of five individuals have yet to be found.

There is only one thing more incredible than all the water and death: in the midst of the rainstorm, clouds were seeded in order to increase precipitation as part of the Bureau of Reclamation's experimental Cloud Catcher project.

The bureau's project sought to build on what initially seemed like a very promising discovery made in 1946. In that year, Vincent Schaefer, a self-trained chemist and high school dropout who worked for General Electric, found that dropping dry ice into some types of clouds caused them to precipitate. That clouds could be modified seemed clear enough but whether, say, a region's precipitation could be increased using cloud seeding was—and remains to this day—much less certain. Undeterred by the lack of statistical support proving the efficacy of cloud seeding, the bureau in 1961, led by the incorrigible river dammer Floyd Dominy, forged its way into the field. Under pressure from congressmen in the arid West, the bureau launched Project Skywater, designed to augment the nation's water resources. Then in 1969, the bureau contracted with the Institute of Atmospheric Sciences at the South Dakota School of Mines and Technology in Rapid City to conduct the Cloud Catcher program, in operation at the time of the 1972 deluge.

Two cloud-seeding missions were flown on the day of the disastrous Rapid City flood. After assessing the day's forecast and detecting no immediate signs of severe weather, the first plane took off at half past two in the afternoon. The plane dropped 350 pounds of Morton salt into clouds located northwest of Rapid City. Before the launch of the second seeding mission, the men in charge of the project, Arnett Dennis and Alex Koscielski, discussed the advisability of sending up a plane given the chance of flooding in the hills area. By this time in the afternoon, the wind had increased to thirty knots. They decided to send the plane up anyway, on a mission south of the city. The weather conditions during the flight were apparently so bad that the pilot had trouble keeping control of the aircraft. Before the plane had even landed, Koscielski was on the phone to the National Weather Service advising it of the possibility of flooding in the Sturgis area, precisely the area seeded during the earlier flight.

Pausing for a moment to reflect, we might say that such foolishness was exactly what the German philosopher Max Horkheimer had in mind in his 1947 book entitled *Eclipse of Reason*. In it he wrote, "The more devices we invent for dominating nature, the more must we serve them if we are to survive." And serve they did indeed that rainy day out in the Black Hills.

To Horkheimer's dictum might be added the corollary that the more we dominate nature, the more natural our natural disasters are made to seem. Soon after the disaster, the governor of South Dakota, Richard Kneip, issued a press release saying that he had been assured that the cloud seeding had not contributed in any way to the flood. According to the release, "Kneip said the last thing that is needed in an emergency such as Rapid City is going through is for unfounded fear or sensationalism concerning a scientific operation that scientists had reported had nothing whatsoever to do with the flood."

The man the governor relied on for assurances that the cloud seeding was not a factor was Dr. Richard Schleusener, a one-time consultant to the Bureau of Reclamation and a major proponent of weather modification. Schleusener was the director of the Institute of Atmospheric Sciences which oversaw Project Cloud Catcher. "I can assure you," he told the governor in no uncertain terms, "that the cloud seeding did not contribute to this disaster." It was, he declared, nothing short of "ridiculous to think that with a few hundred pounds of finely ground table salt disbursed from a single airplane we could cause twelve inches of rain in a few hours." Schleusener's conclusion was as simple as it was emphatic: "There is no evidence that the 1972 flood was other than a natural event." Arnett Dennis, who played a more direct role in the cloud seeding, put an even finer point on the matter. He was dead certain that the seeding had nothing to do with the storm. "I would stake my life on that," he said.

Could the cloud-seeding experiment, which was of course designed to increase precipitation, have played no role whatsoever in the disaster? It was not an easy question and it certainly required a much less categorical answer than the self-serving one provided by the staff at the institute. Fearing a cover-up, Fred Decker, an atmospheric scientist at Oregon State University, wrote Senator George McGovern of South Dakota, urging him to appoint a commission to examine the true causes of the calamity. "Human lives should not be snuffed out because scientists need the results of experiments," he wrote. Decker also explained that he knew from experience "that some of our colleagues who are 'believers' about cloud seeding will hardly provide disinterested or objective advice in the wake of this tragedy."

About that, he was right. Continued public attention forced the governor to appoint an outside team to investigate the role of cloud seeding in the disaster. The investigative team was chosen by Merlin Williams, who was the director of South Dakota's Weather Control Commission, formed in 1953 to regulate weather modification. South Dakota had long been a major center of cloud-seeding activity. In 1951, weather modification was conducted over one-third of the counties in the state. And in 1972, the same year as the flood, South Dakota became America's first state to finance a cloud-seeding program to increase precipitation and mitigate hail. "Control of Nature" ought to be the state's official motto. Of course, were Project Cloud Catcher's seeding missions to be implicated in the Rapid City flood, the prospects for continued state-sponsored weather modification would be jeopardized. In other words, Williams needed to organize a whitewash.

He chose three men to head up the committee: Pierre St.-Amand, Robert Elliott, and Ray Jay Davis. St.-Amand worked at the Naval Weapons Center in China Lake, California, where he helped develop the weather modification technology that at the time was being secretly used to fight the Vietnam War. It came out later in the 1970s that the United States operated a top-secret rainmaking program between 1967 and 1972 that sought to wash out roads and cause landslides, impeding the movement of the North Vietnamese. St.-Amand, a major proponent of weather modification, later testified before Congress about his role in the secret scheme saying that "the main thrust of my professional work has been oriented toward the safe and profitable use of the environment for human benefit."

Then there was Robert Elliott, the president of North American Weather Consultants, a private cloud-seeding firm based in California. Elliott was one of the founders of the Weather Modification Association back in 1951, an industry group formed to promote cloud seeding as a way of solving weather problems. The Rapid City disaster must have seemed a little like déjà vu to him. In 1958, Elliott's firm was sued over cloud seeding it did during 1955 for a California utility company. The seeding was later linked to the devastating Yuba

City flood in which thirty-seven people died. Although the cloud seeders were vindicated in court, the bad publicity generated by the case, Elliott noted, had caused insurance companies to grow queasy over weather modification, dampening enthusiasm for his services. Weather-modified disasters were clearly bad for business.

Ray Davis, the final member of the team, was a lawyer who had carved out a reputation for himself defending commercial cloud seeders. Like St.-Amand and Elliott, Davis also played an active role in the Weather Modification Association, moving on to become its president in 1979. Davis and Elliott had in fact both worked with Schleusener—who in the 1960s led a committee within the association to counter bad publicity over weather modification—to boost the prospects of cloud seeding nationwide. In short, a less impartial committee is hard to fathom.

The three-man panel began its work on June 21, 1972. Just one week later, by virtue of what Williams called "an almost herculean effort," the team released a draft report with the astonishing conclusion that cloud seeding did not contribute to the flood disaster. Floods, they said, occur naturally in Rapid City about once every nine or ten years. A 1907 flood had even swamped an area the same size as that affected in 1972. Their report was an excellent argument in support of clearing the floodplain of housing and letting nature regain control, which, through the provision of federal government funding, is exactly what happened in Rapid City—one of the more intelligent responses to calamity on record in the United States. But keep in mind that the investigators were supposed to be examining the role of cloud seeding in the disaster, not making recommendations on floodplain management. In any case, Williams could not have been more pleased with the committee's work. "I really feel that the whole field of weather modification has been strengthened by the report," he wrote St.-Amand.

The report, which took months to be delivered in final form, drew criticism from around the country. H. Peter Metzger, a biochemist and writer, published a story critical of the committee's investigation in the *Denver Post.* Metzger called the committee's work a "scientific smokescreen" and pointed to some of the ambiguities contained in it. For example, the authors write that their "reasoned conclusion is that the storm was not affected" by the cloud seeding, a point that contradicts the statement, in the very same sentence, that "it cannot be shown that the seeding activities of June 9 augmented or diminished the storm." Metzger's story was later reprinted in the *Los Angeles Times*. Even the *National Tattler* picked up on the Rapid City flood story. "Government Weather Tampering Is Causing World Floods," a huge front-page headline proclaimed. The report cites Metzger as well as an unidentified "independent weather expert." Said the mystery weather man, "We keep telling these guys to cut it out [careless cloud-seeding programs]. The only place we made any headway is with the Rapid City thing. They are scared to death."

Perhaps the most reasoned scientific critique of the cloud seeding at Rapid City came from Jack Reed, a meteorologist who worked at the Sandia Laboratories in New Mexico. Reed published a short letter in the *Bulletin of the American Meteorological Society* in 1973 charging that attempts to disassociate the cloud seeding from the flood were made in haste. A look at the data, he wrote, "does not . . . necessarily lead a critical observer to this conclusion with the *overwhelming confidence demanded* where public safety is involved." Reed argued that one could in fact come up with a theoretical model demonstrating a connection between the seeding missions and the calamity. He concluded that "unacceptable risks *may* have resulted from the cloud-seeding operations."

The letter, really a very tame one, simply suggests the obvious: that the cloud seeding may have contributed to the storm. Regardless of the probability of that connection, this was of course true. Reed did not say categorically that the cloud seeding caused the flood, quite in contrast to the definitiveness of the statements downplaying the connection made by Schleusener and others involved in Project Cloud Catcher. Yet judging from the reaction of Schleusener and his colleagues, you would think that Reed had just cornered the entire market in table salt. In a letter to St.-Amand, Schleusener referred to the Reed "speculations" as contributing to the "'folklore'" surrounding the flood. He also tried to stop the publication of the article, as did Williams. Williams even went so far as to write Reed's employer, the Atomic Energy Commission, to say that "it is a matter of no little concern to me that an employee on the federal payroll should be supported at the taxpayers' expense to prepare and disseminate such information." Reed was careful to say that his views were his own, not those of his employer.

Robert Elliott, asked to comment by the journal on Reed's work, called what he had done an "accusation," which is stretching things a bit. Elliott went on to note that "accusations that cloud seeding has produced disasters are not new. Cloud seeders were most recently accused of causing the drought in northeast United States, and there are many residents of Pennsylvania, Maryland, and West Virginia who are still thoroughly convinced of this," a reference to another battle over weather modification that began in the early 1960s. The weather, he continued, "seems to bear a 'speculative attraction.'" He noted that before cloud seeding, "the atomic bomb was blamed for adverse weather events, and prior to that the introduction of telegraphy was blamed for bad harvest in England." Of course, there did seem to be a much tighter causal relationship between cloud-seeding technology, which was, after all, designed to change the weather, and actual changes in the atmosphere. "The suggestion of holding off on cloud seeding until more is known about it," Elliott concluded, "is no more plausible than the suggestion of eliminating the automobile as a means of transportation until its effects on the environment are fully understood."

St.-Amand also responded to Reed's letter. He said that he agreed with Reed that the public needed to be "protected from irresponsible activity." But existing laws, he believed, were sufficient to safeguard the public. "If every issue becomes a matter of local option—and it may so develop—then no further progress will be made in any branch of technology." St.-Amand was exposing the antidemocratic tendencies raised by the domination of nature, a point that Horkheimer made long ago. As St.-Amand saw it, the people of South Dakota had decided to support cloud seeding, which was true at least in regard to the state-sponsored seeding done under the auspices of the Weather Control Commission. But the experiment in question, Project Cloud Catcher, was not voted on. That did not stop St.-Amand from pointing out that "the public did however approve indirectly in that the Congress, recognizing a possible universal benefit from increased precipitation, allotted money for the work. Moreover, the people of South Dakota approved in that they, through legislative process, decided to let the South Dakota School of Mines and Technology engage in weather-modification research activities." A more desperate attempt to obscure the antidemocratic tendencies of a technological decision to control nature, there has probably never been. For St.-Amand, Schleusener, and the rest of the weather-modification boosters, genuine concerns for democracy and moral reason simply did not enter the picture.

Indeed, the moral issues were largely irrelevant in their crusade to control nature—a mission that, to them, was as natural as it was inevitable. In the 1970s, the Weather Modification Association, whose corporate members included the Los Angeles Department of Water and Power, various electric utilities, as well as private weather consulting companies with names like Atmospherics and Better Weather Incorporated, published an informational pamphlet to educate the public on the "facts" about cloud seeding. Readers were told that there is no such thing as natural weather, at least not anymore. "In the sense that 'natural weather' means cloud formations, storm systems, and precipitation which are unaffected by human beings, there has been no 'natural weather' since the first fire was intentionally ignited by early man." In other words, cloud seeding was as natural as striking a match. Nor was there any need to worry about God or morality. "Is cloud seeding against God's will?" the pamphlet asks. Surely not. "Cloud seeding is a tool to be used by mankind just like a tractor, airplane or space vehicle." Of course to portray cloud seeding as a neutral tool or thing was to obscure the social relations embedded in this technology. Cloud seeding was not simply a tool, but a set of values and moral assumptions about what kind of world should exist and who would benefit from it.

In the summer following the Rapid City disaster, on August 19, 1972, to be exact, someone bombed a trailer filled with cloud-seeding equipment in Monte Vista, in southern Colorado's San Luis Valley. The equipment belonged to Atmospherics Incorporated, a commercial weather-modification company based in Fresno, California. Atmospherics, whose president, Thomas Henderson, traveled the same weather-modification circles as both Schleusener and Williams, had been hired by a group of barley growers to suppress precipitation, and especially hail, during the crucial period in the summer when the crop is harvested. It is essential that the barley not get wet during this time. The San Luis Valley weather-modification program was dreamed up in the 1960s by the Adolph Coors Company, which purchased much of the grain to make its beer. Only ranchers and others in the drought-prone valley objected to the seeding because they felt that it aggravated the dry conditions that at times enveloped the region. In the summer of 1972, a meeting was held to discuss whether Atmospherics should be granted a permit to cloud seed, as was required under a new law passed earlier in the year. Some six hundred people attended the hearing, which lasted until two o'clock the following morning. Most of those attending objected to the cloud seeding, though the permit was still granted, subverting what had promised to be a democratic process and perhaps explaining why someone was driven to violence. Archie Kahan of the Bureau of Reclamation, who attended the meeting, recalled that the loudest applause seemed to follow religious arguments that tampering with the weather was against God's will.

Religious opposition to cloud seeding was by no means uncommon. A survey taken in South Dakota between 1972 and 1974, done by two sociologists, Barbara Farhar and Julia Mewes, discovered that 40 percent of those interviewed held what the authors called a "religio-natural orientation." Asked whether they agreed with the statement, "Cloud seeding probably violates God's plans for man and the weather," two-fifths of the sample agreed. And keep in mind that the survey was taken in a state where there was tremendous support for cloud seeding dating back to the 1950s.

It need hardly be said that Schleusener and his colleagues in the cloud seeding field found the idea that God or nature should be left in control of the weather to be obstructionist if not absurd. Determined to get tough with nature, theirs was a mental universe where the domination of the natural world had been elevated into a moral absolute, where acts of God were relics of a premodern past. Ray Davis, whom we met earlier, put it this way: "I believe that nature is not the image of God; it is his handiwork. In this world, as Eric Hofer [*sic*] puts it, 'anyone who sides with nature against man . . . ought to have his head examined.'"

Also weighing in for *Homo sapiens* was Edward Morris, like Davis, a lawyer and president in the 1960s of the Weather Modification Association. Morris had successfully defended a California electric company against charges that its cloud seeding had caused a flood to occur. "Whether or not we all like it," he wrote, "large-scale weather modification lies in man's future." Seeking to further naturalize attempts to control the weather, Morris pointed out that "unintentional" weather modification through industrial activity and the release of carbon dioxide and other pollutants was already taking place and had been for some time. With weather modification already occurring, albeit unintentionally, what harm was there in purposefully trying to alter the weather to the people's advantage? The only problem was what to do with those opposed to cloud seeding. "To overcome the emotional opponents of weather modification," Morris explained, "it will require more than a letter from a government agency that its files contain no evidence of disasters caused by cloud seeding." It was going to take a federal agency organized along the lines of the Atomic Energy Commission or the Federal Communications Commission, Morris explained. "This Federal agency might eventually plan weather by zones or by days, sell or buy weather to or from other agencies, possibly even make decisions as to the 'best' weather for certain times and places."

Such insane notions notwithstanding, the cloud seeders continued to see those who stood in their way as lunatics. Schleusener himself even went so far as to compile a file on those he viewed as crazy enough to oppose to weather modification. He labeled the dossier very simply "Crank." Admittedly, the crank file does contain some real oddities. Francis N. Bosco of Lakewood, Colorado, who fashioned himself a "weather engineer" and a "practical meteorologist," tells of a plan to solve Denver's smog and unemployment problems all at once by burning some secret chemical and blowing it into the atmosphere. It was not weather modification per se that Bosco objected to, but the failure of what he called the "un-ethical group at Rapid City . . . to consult my advanced procedures."

There are, however, materials in the file written by people with legitimate and honest religious and moral objections to the idea of weather control. Mrs. Adolph Hermann of Lemmon, South Dakota, writes to complain about a vicious hailstorm in 1968 that was apparently seeded. "Seems it would be much better to let the Almighty God handle the weather situation," she writes.

Gertrude Milton Walcher of Colorado Springs also wound up in Schleusener's filing cabinet. In a pamphlet titled *How to Kill a Beautiful World,* Walcher criticizes cloud seeding and other attempts to, in her view, destroy the earth. "This thesis is being written because of the righteous indignation, anger and frustration of a subjective pioneer in a world obsessed by objectivity and materialism," she begins. "It was a beautiful world when the laws of God and Nature were revered," she laments. "Now it seems that there is nothing that is sacred."

Clearly Schleusener did not take what Walcher had to say seriously. But we should, because she offers the start of a useful critique of the cloud seeders. To the weather modifiers, she points out, nature has no intrinsic meaning or moral value. "One would think that they would have some compunction about destroying God's handiwork, but evidently such an idea has no meaning to them," she writes. For the cloud-seeding boosters, nature is little more than a resource, with no meaning outside its value as an instrument in human economic development. Moreover, those seeking to obstruct the domination of nature were seen by the seeders as premodern in their thinking. Opponents of weather modification were made out to be weak, superstitious, and emotional, and too prone to accept acts of God or nature when they arose. And yet, it is also clear that the natural-weather people were very threatening to Schleusener and his weather-modified world. The cranks showed that cloud seeding was not normal or inevitable but something that weather entrepreneurs, government bureaucrats, and political leaders were foisting on unwilling sectors of the American public. Above all, the cranks were dangerous because they interfered with the attempt to naturalize natural disaster.

When weather-modified calamities occurred, the seeders argued that larger natural forces drowned out the intended effects of the seeding. But the cranks had an uncanny way of taking seriously the efficacy of weather modification at precisely the wrong time. They sought to inject moral responsibility into the debate over natural calamity and thus earned the enmity of the cloud seeders who tried to block such ideas. They caught the weather modifiers in the thick of their contradictions, showing them to be not only oblivious to the role of human responsibility and morality in disaster, but at times illogical—their common sense clouded by a technocratic dream to control the uncontrollable completely. To this day, statistical evidence supporting the efficacy of weather modification is virtually nonexistent.

Which leads me to one final suggestion: The next time you find yourself driving the lonely stretch of Interstate 90 in western South Dakota, stop in at the School of Mines and Technology in Rapid City. Go to the library and ask for Schleusener's crank file. Read it and the rest of his papers and then ask yourself: Who's really crazy here?

# III

## *Diving Deep*

Marie Wilkinson/Cyril Christo, *Patagonia.*

# A Myth for Sofie

*C. L. Rawlins*

Like injury and disease, science was something I avoided. It wasn't a lack of interest, since I could spend hours on my knees to watch how a horned lizard devoured ants, or how the watery glaze left a freshly poured sidewalk, which gave off a gentle heat as the concrete firmed and set.

This passion for discovering how things work has never lessened. So the reason I hated science, or the one that occurs to me now, was the way in which it was taught. In the schools where I served my time, it was taught in a way almost perverse, not as a means of appreciating the natural world, but as a denial of it. Striking events—thunderstorms and lightning—were reduced to fat blocks of text with a few smudged photos, followed by Key Concepts. For instance, it takes a charge of about 300 kV per meter for a spark to propagate through air. Gee whiz.

On rare days we might watch a film. At that time, science films were mostly black and white and narrated by former insurance salesmen. What we never did, even when thunder echoed from the peaks to the west and the air smelled of rain, was to go outside. The purpose of our instruction was to portray the world not as exciting or yet mysterious, but as raw material. It wasn't worth a great deal, this world stuff, until science had transmuted it into a new floor wax or explosive warhead—or, at the very least, reduced it to numbers and stacked those numbers into walls of text. Given a choice between dull obsession and the world, I chose the world.

The college science I had to take was suffered as I partook hungrily from the other end of the board: poetry, sculpture, photography, history, anthropology, and linguistics. Between bouts with the arts and humanities I worked outdoors. In the Forest Service, my skill at packing horses and locating camps got me involved with scientists who required such help, and eventually I found myself collecting samples—rain, snow, lake water, plankton, mayflies. I liked it because we were testing real things, dipping bottles into snowmelt creeks and feeling the cold in our bones.

The Forest Service is hard on real scientists. They arrive with high hopes and last a season or two, then go back to school. So before long I was running the field project without the proper education. Outdoors, I'd learned some rules by heart—gravity, for instance—but that knowledge wasn't certified. And I still felt a poet's distrust: science had penetrated the natural mysteries and used them to manipulate and deform. Worst of all, it might dispel the wonder that coursed, quick as a bolt, between the world and me.

But I loved the work and wanted to keep the job. So in 1987, at Utah State, I enrolled in the classes I'd avoided: brute chemistry and physics, followed by watershed science, hydrology, aquatic ecology, and limnology.

What's limnology? The study of freshwater streams and lakes. And from those bodies of water, I'd gathered questions that couldn't be answered in poetic terms. For instance, why did long, straight lines of bubbles form on alpine lakes? In a marsh, bubbles arise from decomposition. But our lakes are set in granite, cold and dilute, with too little organic matter in their beds to yield much in the way of methane gas. Froth can also form by wave action, but these parallel tracks appeared when the wind lacked sufficient force to churn air into the surface. And I'd seen them, evident on the dark water as the Milky Way on a black, alpine sky.

The limnology course was stocked with broad-beamed guys in ball caps: fisheries majors, whom hydrologists call "Fish-Heads." Being graduate students, they'd shared the same classes for a year or two, and they viewed me with suspicion. And gazed likewise, with knitted brows, on the woman in the center of the room.

She didn't have that stolid, Fish-Head look. Instead, she had a heart-shaped face and a piscivorous glance, quick and disconcerting, like a mink. But she was far from sleek. Her red hair was joyfully disheveled, gathered once at the crown and then let fly, and she wore a gray sweater so ratty it looked like real rat pelts stitched together, three sizes too large. Beneath, in plum tights, stretched dancer's thighs and calves, which funneled down into green-striped socks, bunched above surplus boots.

She wasn't beautiful, exactly—she was stunning.

So. I tipped my head toward the empty stool on her left, and she nodded and shot me a smile that as soon as it registered was gone.

We whispered. Her name was Sofia Wikberg, a Swede, and a vicar's daughter. But what was a vicar? I remembered those grim clerics in Bergman films, squinting across acres of lace tablecloth. But she didn't squint. Her blue eyes flickered through a net of rose-gold hair. *Sofia* means "wisdom," of course, and she seemed wise. Despite her punk-ballerina attire she was reserved and precise. An aquatic biologist, she had a fiancé who was a doctoral student in chemistry; they would return to Sweden when he finished up in May. Until then she was marking time.

We claimed adjacent stools in the second row. I began to come early, ostensibly to study, hoping to talk. In the second week she did too. When I greeted her as Sofia, she said, "Oh, Sofie is good now, I think," and gave me a fleeting smile. Dismissing the class as "Oh, too easy, isn't it?" she cheered me with Euro-cynicism as she helped me comprehend the math. Even so, the five sessions per week seemed hard to me, but on the first test—lake morphometry—I got a 98, and she got a perfect score. But my lack of math began to tell, and by mid-quarter Sofie carried the only A. I had a desperate B. The Fish-Heads, sunk in C's and D's, began to grumble as they lumbered in each day to find the two of us conspiring.

For the most part she ignored them simply, but she could also project a blazing disregard. At times, when I asked about an equation, she'd explain in breathy whispers and then talk, in a voice that carried, about my poems:

"So you've published *po-et-ry?* Oh, a *book?*"

Did she write poetry?

"In grammar school I did. It was not good."

Would she like to read my book?

"Well, *perhaps*. Poetry is so *in-ter-es-ting*."

The Fish-Heads flinched at that, and I expected one of them to launch a limnology text at my skull. It was their class, and we were spoiling it, the poet and the smart-ass foreigner. It wasn't fair.

Plato, that ancient enemy of poets, believed that knowledge comes to us in four ways: thought, feeling, sensation, and intuition. Thought is the measured process that yields equations. Feeling lends us our values and prompts the judgments that arise. Sensation is weight, depth, texture, heat, the darkness, and the light. And intuition is the gift of seeing to the heart of things.

We combine these faculties, each in different proportions, and lacking one may grow strong in others: a scientist in thought and a revolutionary, perhaps, in feeling. But intuition is particularly the poet's gift. For a poet encounters the same rocky substance and lives by the same laws, gravity, and death, yet offers it up as a potent music.

Sofie, the scientist in her woman's body, pursued rationality with fierce emotion. She delighted in the cool joinery of equations, and could wither with a glance. Yet to me the numbers could never bear the weight of real things. *Velocity* wasn't just a term in an equation, but a cold creek tugging at my hand. The world itself was lodged in me, and science couldn't displace it.

My gifts are otherwise, sensuous and intuitive, and each new trick I learned was full of curious potential. Besides learning to work the equations, I collected words—*aphotic, hypolimnion, profundal*. But with marriage, classes, and a part-time job, I lacked much time to write. So limnology class was the high point of my day, as I sailed in Sofie's bright mathematical wake, and as our aquaintance ripened she took on a rosy light. I'd met her fiancé, and she knew that I was married. She'd admired my ring, a copy of a museum piece, like two golden twists of rope. I slipped it off, and she held it up, then looked through it with a bright blue gaze that reminded me of high lakes in the afternoon, and of their strange and parallel effervescence.

Why oh why *did* long, straight lines of bubbles form on alpine lakes? In lakes set high in granite, cold and dilute, with too little organic matter in their beds to yield much in the way of methane gas. Froth is formed by wave action, but these tracks appeared when the wind lacked the force to churn air into the dark water. And I'd seen them not once but many times, plain as plain could be.

And halfway through the text, I found the answer: "Langmuir (1938) demonstrated that under some circumstances the motions associated with turbulent transport are organized into vertical helical currents in the upper layers of lakes. Convection from this vertical motion generates *streaks,* which are oriented approximately parallel to the direction of the wind. Streaks coincide with lines of surface covergence and downward movement. . . . Between the streaks are zones of upwelling." And bubbles, by God! Ecstasy.

Sofie arched her coppery brows. I touched the diagram and made frantic gyres.

"Oh!" she said. "The equations are *very* nice."

And suddenly, I could appreciate that they might be. And on the heels of that, I knew that science couldn't disarm my wonder. Faced with something like, "The Langmuir convectional helices form a series of parallel clockwise and counterclockwise rotations," I could see dark lakes and mysterious bubbles.

Sofie had given me the gift of numbers, with no false promises. I wanted to do a corresponding thing for her, so I started to write her a tale. I was trying to account, in a poet's way, for those bubbles, those wonderful helices, and distill for her a myth. But the quarter ended in a flurry of tests and papers, and the myth was left undone.

Afterward, I saw her seldom: in the halls of the Natural Resources Building where we exchanged quick grins, or passing with opposite strides on a walk. I saw her one last time on a ski trip high in the Bear River Range. After skiing the crest in a storm, we descended to check on two friends, Keith and Tracy, at their camp. Tracy had a trick knee and had strained it. We asked whether she wanted us to bring up a sled and ease her out, but she thought her knee would hold. The storm had abated, and the fresh snow was perfect for turns. We were talking when Sofie skied out of a cloud, hair escaping a frayed cap, followed by her dark-headed chemist.

"We heard voices and thought—Oh, it is you!"

She was all there. Then she was gone. In this world there's far more beauty than we know, and more ways to find it out. Each time the breeze comes up and leaves shimmering tracks on the lake, I think of her. And now that it's finally done, I want to give her this:

### *A Myth for Sofie*

*You've heard of the Lorelei, those water spirits whose songs lured so many boatmen to drowning. This is a story not about them, but about their younger sisters, the kinder ones. Of them you have not heard, who fled the strong currents of the great rivers, and the rush and boom of big boats, and the hoarse cries of boatmen.*

*Where have they gone?*

*I found them. I haven't seen them, of course, since that is a more difficult thing.*

*But I've seen their songs.*

*What, you say? How does a song look?*

*Be patient. We will talk first of the gentle sisters, and how they left the great, brown rivers of the lowlands. How they were tired of the mud and silt, and weary of the endless traffic of boats and barges. How they were saddened by the cruelty of their older sisters, the angry ones, who knew the beauty of their songs and used them so ill. And how they were sickened, those shy and quiet spirits, by the cries of drowning men.*

*Not wishing to surrender their fates to the current, nor wanting to follow the bodies of drowned men drifting toward the sea, the younger sisters fled upstream. Where two rivers joined, they chose the clearer one, and followed it. In time they left the broad rivers for the rolling streams of the hills, and rested there, singing softly under the willowed bends.*

*But one day a child wandered away from a village and came to the stream. The child heard their quiet songs, saw bubbles in the current, bent down from the bank and leaned too far, and fell into the water.*

*The sisters of the Lorelei watched the child struggling in the water and could not save it, for they themselves were water. Their hands were water, and they tried but could not lift the small body free of the water's grasp. So the child drowned.*

*Singing in whispers, they mourned. And then they left the lands of men.*

*At the top of the world is a lake. There the rain falls free and the snow falls and melts without a track, and the waters divide, to flow down to all the inhabited lands. And there, on top of the world, the water is clean and clear and cold.*

*The lake is so deep that the sun cannot penetrate its depth. You can, if you choose to risk, climb the peak that rises above the lake and look down and let your gaze go into the clear water. You may climb to the very top and rest in snow and ice—if you go that far—and try your best to see the bottom of the lake. Under all water, no matter how deep, there must be rock. But you will not see it, except at the very edge as a narrow ring of jade and turquoise. Even at noon on the highest day of summer, the lake is a great, black jewel.*

*And there, in the deepest part, is where the gentle sisters have chosen to live and where, unseen and unseeable, they pass earth's time. In the depths they sing and cannot be heard, at least by human ears. So the beauty of their song is no longer a threat.*

*From the transparent darkness they watch ice form above, a thousand blues, and watch it melt. They watch trout swim far, far above, like curved stars. In the brief alpine summer, the surface is a shimmering blue, like a canopy of silk. And when the lake is riffed by the wind, they join hands and sing.*

*Under the lake at the top of the world, the sisters join pale hands. To the water's gentle beat, they form long lines to dance. And from their lips rises a song.*

*Yet the listener, if there is one, hears nothing. Because there is nothing to be heard, in the air of this world, but the rush of wind and the slap of small waves on the rocky shore. And the watcher sees only a long line of bubbles, wavering up into the light. Have you ever seen a song?*

*Perhaps you have. For now you know they are songs, these bubbles rising from the lost, dark lake. They are songs which no living person may hear, and still live.*

*Oh, for such beauty to exist and be withheld. The world's unkind, you say.*

*And do you complain of the stars, that we cannot feel their heat?*

*Must stars sing too? Why, when they breathe perfection, silent on a clear, black sky? And if they were to sing, each star in its own, wild, burning voice, how could we sleep?*

*So the world is kinder than we think.*

*And so it is with the sisters of the Lorelei. And so it is with their music, those long traces on the windy lake, each song true and distant, quiet as the Milky Way.*

## Note

The scientific quotations are from Robert G. Wetzel, *Limnology* (Philadelphia: Saunders, 1983).

# Agate Beach in Rain

*Laynie Browne*

Marilyn Ulvaeus, *Rice paddies in Bali.*

Agate beach in rain which fell heavily, creating a
pattern of opening and closing cavities of sand. The
sound guided further out to sea. A blanket thrown
back to a warm December. Hibernation in winter is
the same misrepresentation of venture; if I remain
indoors I may travel more easily. Speech mirrors
silence the way footsteps would otherwise remain
submerged. Ambulation is changed by water.
I travel away in order to match the hem of waters,
undulations around ankles. I witness the tide's
undermarkings, as if reading a series of utterances.
An attempt to study the kaleidoscopic languages.
A seal swims in pink waters. The red tree is certain
to find me. His helpless destinations became
nowhere in mind, where luck pronounces soil.
Forbidden scent of floating gardens. Floralia.
Sabrina's early haunt on the bosom of waters.
Garlands put to sea. His hands became extensions of
the landscapes he created, hovering forms. An
island once underwater now awaits the return of
trees.

# Blue Moon Tide

*Kartik Shanker*

## *Prelude*

*That night in San Francisco the moon was blue, I am sure of it. It was a wonderland late evening, a neverneverland night, with stars twinkling like notes of music plucked from guitars and the notes plucked from three guitars lit up the evening like a skyful of stars. Some dwindled, some lingered, some exploded like supernovas, leaving black holes in our hearts. The golden gate hummed in the background, lights glowed in drawing rooms, dining rooms, bedrooms, people talked, ate, fucked. But the space around the guitarnotes swelled into a niche in the universe that was all its own, the tunes stretched the seam of our lives, poked at the fabric and demanded to let us out of our pathetic skins, and burst from our pores and swung and danced on the tips of our fingers, delicate as ballerinas, joyous as Indians, whooped, and pirouetted, in harmony and synchrony, and joined us all in a faultless seam, pierced our hearts with guitarstring, wove a pattern of enchantment that we were sure was love. And went on and on until it was like rain falling heavier and heavier until our hardly solid bodies, unable to resist the torrent, dissolved, melted, drained away into rivulets, collected other little fingers of water and bubbled into larger puddles, swelling with the blue moon tide, with heaves and sighs of content, lapped upon the shores, beat upon the shores, ripples, currents, all together and separate, waves crashed like fingers running down a piano, again and again.*

### *Arribada*

Many years and light-years away, on a long lonely strip of beach, a lone olive ridley sea turtle clambers onto the beach, leaving the tracks of her body and flippers on wet and dry sand, an ungainly woman climbing a horizontal ladder. The dry sand sticks to her wet underside—a pale plastron, and to her fore and hind flippers, to her neck which she lifts periodically to breathe and to survey this stretch of land, so strange and yet so familiar, home and not home. Finally, avoiding the long creepers that we call "goat's foot," *Ipomea,* she clears a little space for herself in the sand, stretches a little hollow to hold her body, pokes out the sand that resists. Then, with precise circular movements of her hind flippers, she carves out a nest in the sand below. And when she scoops the sand with her hind flippers, she always alternates flippers, never uses the same one twice in succession. Even when one of them is mutilated, or deformed, or bitten off by a shark, she moves the stump, trying to scoop sand with her phantom flipper. Finally it is ready, a subterranean dwelling for her eggs, a flask-shaped hole that is about a foot wide and a foot tall, with a narrow neck leading up to her. Once this is done, she goes into a trance where nothing exists except her and the nest into which she lays her eggs, two and three at a time, small and white like Ping-Pong balls, but soft, dropping them into the wet earth below exactly as she has been doing for millions of years, long before man was even a twinkle in the eye of natural selection.

*. . . I have been around even longer than the turtles, longer than the little hatchlings I have carried from shore to seaweed and later from seaweed to shore, longer than the beaches, and much much longer than the toes whose interstices I fill with sand . . .*

The eggs fall with mucous wet plops into my waiting hands below. I am crouching behind the turtle on my hands and knees, I have dug a small hole behind her after she started laying her eggs and I collect them into a cloth bag while Jo holds a torch for me. She is sitting on her knees beside the turtle and giving her moral support:
"Come on sweetheart, you can do it, just another hundred eggs," she says, moving to sit near the turtle's head. "Turtle tears, she's crying for hatchlings she will never see."

"Don't get soppy," I say, though I know that there is very little about Jo that is soppy. "It has to do with salt balance when they're out of the sea."
"I know."

The turtle stretches her neck and strains, and another few eggs fall into my hands. Jo flashes the torch at the turtle's face, something sure to send the turtle scrambling back to the water at any other time, but not now that she is in her nesting trance. She reaches out and examines the turtle's flipper and runs her hand down the tendons stretching in her neck, and says,
"She looks as if she doesn't even realize we're there. Which is remarkable considering they're so skittish before they start to lay."
"Remarkable," I say, trying desperately to keep count of the eggs. Thirty-four? Forty-three?
"Absolutely. I suppose you could set her ass on fire and you wouldn't disturb her now."

The girl has a way with words. I collect the rest of the eggs from the nest and place them carefully in the bag. They will be relocated in a hatchery, a rectangular length of fence that hopes to keep out intruders by staring them down rather than posing a physically impossible barrier. This stretch of beach has been populated more and more in recent times; dogs and jackals have depleted the turtle population by feasting on the eggs, and humans have also helped to harvest a meager crop. So it is left for a few volunteers to stalk the beaches and steal the eggs from beneath the belly of the turtles and take them to safety. A hundred-odd eggs later, she is finished, and we watch as the unsuspecting turtle fills up an empty nest, beats down the sand with laborious heaves of her body, flings sand around in a million directions, circling on her belly, then pauses to figure which way the sea is and scramble-lumbers back to a waiting tide.
Jo says, "We could eat the eggs, and she would never know."
Oh, but we do.

At the hatchery, I dig a pit in silence. Jo has deserted her lamppost duty, her lady-with-the-lamp act, and the torch is perched on a post some few feet behind me and lights up a patch of sand ten feet away. But I do not need to see the sand any longer. Like the turtle, I only need to feel it, against my fingers and bones, beneath my nails and skin, as I scoop out a hollow, as like her as I can manage. I count the eggs into the nest, 147 this time, thinking as usual that only one in a thousand hatchlings survives to adulthood. After I have finished relocating the eggs, I find Jo at the water and hold her hand while we let the sea push sand between our toes.

There are seven species of sea turtles, including the enormous leatherbacks, so named for their cartilaginous carapace, seven feet in length and weighing several hundred kilograms. And then there are green turtles, which have been hunted for their meat for centuries, the turtles of turtle soup fame. And hawksbills, with their beautiful scutes, peeled off their backs (sometimes when they are still alive) for quality tortoiseshell glasses and combs. And there are loggerheads and Australian flatbacks. And of course there are the ridleys, the smallest of the sea turtles, less than three feet long, mostly less than 50 kilograms. The Kemp's ridley, a native of Mexico. And the olive ridley, a citizen of the world. Found in all the oceans, from the beaches of Mexico to the sands of Orissa.

Bhitarkanika is a delightful sanctuary on India's Orissa coast, with a lush mangrove forest and an assortment of interesting animals, including the saltwater crocodile, or salty, and the king cobra, neither of which ranks as the Wildlife Channel's no. 1 cuddlies. At the mouth of the River Maipura, which runs through Bhitarkanika, lies Gahirmatha, the original site of the world's largest sea turtle rookery. Half a million turtles used to nest on a ten-kilometer spit in the short span of a couple of weeks each winter. In 1989, however, a cyclonic storm bit off the end of the beach like the butt of a cigar and spat it into the sea. The olive ridleys continued to nest on this four-kilometer-long, hundred-meter-wide island. In 1997, the island split into two, leaving a northern portion that was just two kilometers long and less than fifty meters wide in places. The ridleys continue to be faithful to this patch of moving land, this island that is shrinking each year and floating away to the north. How far will they follow it? Where will they go when the island is gone? Will they, finding that their incubators are gone, find secret shores to entrust their eggs for safe keeping? Will they instead wander forever, chasing a magnetic memory of a beach that is not there anymore?

Jo pulled her hand away and dusted the sand off and put her hands on her hips and looked across the sea. I looked across the space between us, tight

jeans rolled up against a soft calf, loose sleeves flapped against tight shoulders, a familiar faraway expression floating across her face. Jo had inherited that. I reached for her waist and shook her.
"Hey Jo, what's the matter?"

She looked at the highlights in the water, *Noctiluca,* microscopic algae that luminesced every time the waves crashed them upon the shore, each wave appearing like a phosphorescent radioactive cloud, wrapping around our feet like glow-in-the-dark condoms, leaving green-glowing specks on our heels and toes, and said,
"Unruhe. Restless. I'm restless again. I do have to go back . . . don't I ?"
I shrug.
"Maybe. If that's what you want. Anyway, wait until the end of the season. You have to see an *arribada.* And of course, hatchlings."
"I'm really happy I came here. With you. I don't feel out of place anymore. But it's like, you know, the holiday is over and it's time to get back to work. Of course, I still don't know where I'm going."
"You'll find your nesting beach one day."

Her eyes twinkled for a moment, like a wave had crashed the *Noctiluca* in her eyes, and she put her arm around my waist and said,
"If the trawl nets and the sharks and killer whales and other monsters don't get me before that."

In the North, the turtles are arriving, in tens and twenties and then hundreds, thousands, hundreds of thousands. The males are already on their way home, but hundreds of thousands of female turtles swim offshore, waiting for a cue, for a clue. The sea is crowded with half a million waiting, expecting, expectant turtles.

Only the ridleys do this. All other sea turtles nest solitarily. The wandering leatherbacks, black and enormous, who dive 4,000 feet in search of jellyfish and swim into the open seas following oceanic ridges, nest solitarily. Green turtles, swimming halfway across the Atlantic to nest on a speck of land called Ascension, nest alone too. And hawksbills, wandering among sunlit corals, nest by themselves. And loggerheads and flatbacks. But ridleys . . .

On a moonlit or moonless night, they all decide to come ashore. *Arribada.* The arrival. In Gahirmatha, the turtles prefer the half-moon—not too full, when it is neither too bright nor too dark, when the waves are not too high or heavy, neither spring nor neap tide, and the south wind is rushing over the island and fine sand screaming across the beach. It is almost as if the blistering wind blows the turtles onto the beach. And so they come, wave after wave of turtle, hundreds are swept ashore, and then they lumber up and wander amid the sea of turtles already on the beach, looking for a place to nest. Some choose to return and come again a day or two later. Others find a site that suits them and clear a patch, throwing sand onto other turtles in the vicinity and then, while they are laying, are covered almost completely by the sand thrown by other turtles. And so they keep arriving, nesting then leaving, nesting and leaving. The beach is thick with turtles pushing, shoving, digging their nests, digging the nests of earlier turtles, scattering their eggs to the winds, eggs flying, eggs laying, all through the night. And then as the sun rises, the last of the turtles wanders back to the sea, and it is over. In the morning the beach seems as placid as the sea. Who would believe there are a million eggs buried in the belly of the beach. For a week or two, they come ashore every night, sometimes as many as fifty thousand, all nesting on a patch of sand that is being blown away to sea by an intractable southern wind.

*Amartya and I read about the turtles together. We learned that only the Kemp's ridley nested during the day, but that afternoon, as we chugged in our little boat toward the nesting beach, we could see sand flying over the tops of the dunes, and as we came closer, saw a few turtles, flapping around, clearing areas for their nests. We jumped ashore and ran to the top of the dunes, and as we came over the top, stopped. For there in front of us, as sudden as a storm in summer, there were five thousand turtles on the beach, all crawling around, looking for a place to nest, while others, placid as rocks, were already nesting, and at the water's edge, each wave washed ashore dozens of females that began crawling hurriedly up the slope. As the sun set, the beach turned a golden brown and the features of the turtles blurred and turned into silhouettes in the moonlight. Later at night . . . there were fucking turtles everywhere, crawling on the sand like leeches, worms, maggots in a carcass, vultures at a corpse. Like geese, gulls, petrels they flocked ashore. Like soldiers in a battalion, they marched ashore, only their helmets visible over the ground, thousands of buried soldiers on the beach, the round domes of their helmets sticking above the ground catching the moonlight and flicking it around at each other. Like ships in a fleet, like tanks, millions of little mounds in an armored division. They flooded ashore and lay there gasping like a school of beached fish. They bumped into each other, crawled over each other, threw sand into the eyes of nesting turtles, dug out the nests of their sisters, careened off each other in a frenzy, like the Disneyland bumping cars in slow motion. Like flies, maggots, bees, locusts, passenger pigeons, they kept coming and coming, comical in their labored insistence. And some were laying even before they could dig their nests, as if they could contain themselves no longer, a kind of reproductive diarrhea. Two turtles nested back to back like bookends, two turtles nested on top of each other, like a sculpture of turtles mating, two turtles were surrounded by the shells of the eggs they had dug out. And so many were covered up to their necks in the sand thrown by others around them, they looked dead until I pushed at them with my feet. Hell, I can try until the moon and I are blue, but I can't describe it.*

### Juvenile Frenzy

About seven weeks, the eggs have been cooking in their earthen oven, warmed by each other and the sun. The warmer ones will become females, the colder ones males. Is it ever so? Crocodiles, on the other hand, develop into males at warmer temperatures and females at colder incubation temperatures. But then they shed crocodile tears. Turtles cry for the eggs that they will leave and never return to. And what of humans? Ah, sahib, we are turtles all the way down.

Finally, the hatchlings are ready to face the world, and they chip and crack the shell with their caruncle, the egg tooth, a keratinous saw at the tip of their snout, and then push and prise away until they divest themselves of their calcareous wombs. Then they wait until the dark and until a fairly large number of their siblings are ready to make the break for freedom with them. Night falls, and the cool night air blows over the nest, cueing the hatchlings beneath. And then they all start moving together and the sand moves with the bodies of a hundred hatchlings, as they all rise, upwardly mobile, and the nest caves in behind them, until finally they emerge, feel the spray, smell the sea, see a brighter horizon. And take off in a wild race to get to the sea, frantically scrambling over the sand, each human footprint a valley and every minute dune a mountain, and they tumble down the final slope to brave the waiting breakers. Some do not reach; they are intercepted by dogs and jackals and

Bivash Pandav

birds and crabs. But as each hatchling reaches the sea, feels the first wave wash over it, dives under the breakers, swims against the incoming wave, with the outgoing one, doggedly maintaining direction, and finally makes it out into the open sea, it knows that this is where it belongs. The juvenile frenzy is over, the hatchling's reserves of food have been used up getting it out here into the open sea, and it is time to start living the life of a little nomad. Now it is subject to the vagaries of currents, tides, inclement weather, and waiting mouths.

Jo lifted a basketful of hatchlings that had emerged in our hatchery and walked towards the shore to release them. As they scrambled back, I said,
"Isn't it marvelous that they know exactly which way to go?"
"They're attracted to the light," she said, bending to flip over a hatchling that had been upturned by an indifferent wave.

After emergence, hatchlings move toward a brighter horizon. This is invariably the sea, because natural light from the stars and the moon will reflect off the water. Add to that the silhouettes of dunes or trees behind beaches, and the sea beckons brightly to the hatchlings as soon as they emerge from the nest. And as soon as they hit the sea, they swim against the waves, and this is when they get oriented to the earth's magnetic field.

*But when there are all kinds of other lights—tube lights and streetlights, and the glow of a smoking city, which lingers like the tip of a cigarette—they get confused and disoriented, head toward the streets, the highways, get run over by trucks carrying sand and cars carrying happy families or die of dessication in the morning. Late-night city lights can have pretty much the same effect on Amartya and me.*

Jo was conceived on a music-lit December evening in San Francisco. I bumped—literally—into a woman carrying a Coke and a burger, and we became acquainted over cola stains on her dress and a mess on the floor. I bought her another snack, and later we held hands and got lost in the evening haze. She was long and slender, gawky and graceful simultaneously, somewhat like a flamingo, pink too in parts. She had also the most uncontrollable hair that I have ever seen, a face that was all angles and accidents, thin lips that belied a gorgeous but rare smile, a beguiling nose and spaced-out eyes. That evening I put it down to marijuana, which is why I looked the way I did. But I discovered that spaced-out eyes were a permanent feature with her. I was then twenty and feeling for the first time a nagging restlessness, the migration urge, a terrible wanderlust. I had lived all my life in Bombay, where my father, Subhas Dash, formerly of Cuttack, Orissa, fighting the lethargy that often afflicted those from that part of the country, had more or less succeeded in making his fortune as an accountant in a nondescript firm. My siblings and I became doctors, engineers, and lawyers. They got married, had kids, were mildly corrupt, worked eighteen hours a day, and moved gradually through larger apartments and bungalows and increasing numbers of cars, toward a nirvana they believed was an eventual orgasm correlated with the size of the extended penis. I thought there was more to life than that. There was, of course, America. America aah-mer-ica.

When I first braved the breakers, and swam beyond into an ocean I neither knew nor understood, I was completely lost. The currents carried me across the continents through a featureless ocean, and I scrambled for shelter to the first seaweed raft I saw. Gina and I shared an apartment that amply reflected our internal chaos. She did not understand how the same smells, the same smog, that was her warm security blanket was my dark cloud of depression. I constantly paced the room, searching for a way out, looking for escape velocity, a critical threshold. The sofa in the center of the room was our presiding deity, adorned with my jeans, her underwear, our computer, her music, my notes, and an assortment of junk that I did not dare investigate for fear of unleashing toxic fumes. A TV sulked in one corner, a refrigerator glowered in the other, and the stove simmered between, completing the menagerie of bad-tempered animals. Gina was mostly silent and watched the distance fade slowly, as she did when she was irritated or distracted. Her fingers fluttered like something was just beyond her grasp. We went on with each other for a while, playing chess with each other's emotions, the queen's gambit, the Indian defense—things like that. Of course it would end in stalemate. Gina resolved that by kicking free of our weedhouse and swimming south that winter. I remember the time she left, the room was lit by a purple haze, which may have been grass or alcohol or smog seeping in through the corners of the evening, and everything was sharp and distinct, including Gina. She seemed weary, and I did not have the heart or the energy to push or to pull. The tides do a good enough job. So I held her hand and let the sea push sand into the interstices of our toes, knelt in front of her. She sat stiffly on the sofa like a monarch about to abdicate, a young girl about to confess her first sin, a lover about to leave. She left shortly after, taking with her a daughter who was still outgrowing gills in her uterus.

### *Interlude*

"Have you heard that story about turtles?"
"Which one?"
"Well, a native is telling a pink caucasian that there is this legend that the world rests on the back of an elephant. And what does the elephant rest on? freckleface wants to know. On a turtle, the native says, raising his eyebrows with astonishment at such a silly question. And what does the turtle rest on, the intrepid phirungee boldly enquires. Ah, sahib, it is turtles all the way down."
"Ah, sahib, that one."

There is this story in Hindu mythology. The Devas (the good gods) and the Asuras (the demons) were having one of their numerous skirmishes. On this particular occasion, the Asuras, armed with some newly acquired weapons technology, were beating the pants off the Devas. The good gods, as usual with their tail between their legs, went to Big Brother Lord Vishnu for help. Vishnu advised them to seek *amrita,* the cream of the sea of milk, which would give them strength and immortality. To get this job done, a matter requiring no mean engineering competence, they had to churn the sea with the Mount Mandara, using the giant snake Vasuki as a rope. Of course, the Devas had to co-opt the Asuras into the project, as it was too large for them to undertake by themselves. Despite their best efforts, though, the mountain kept sinking. So Big Brother Vishnu turned up as a turtle (though it is translated as tortoise, it must have been a turtle, and a sea turtle at that, making its first appearance in literature, though the species is not specified) and held up the mountain while the minor gods churned away.

"Of course, Big Brother had no legal, moral, ethical, or political standing to interfere in the internal affairs of another government."
"Actually, Vishnu made rather a habit of turning up in one fancy dress costume

or the other to bail out the good guys. He had already been down as a fish and would later appear as boar, half-man/half-lion, priest, warrior, king, and lover. Apart from amrita, the sea also regurgitated Dhanwantri, the keeper of the cup of amrita and the gods' physician, Lakshmi the goddess of wealth and beauty and wife of Vishnu, Sura the goddess of wine, Rambha the nymph, Chandra the moon, Parijata the celestial tree of wishes, Kaustubha a jewel, Dhanus a bow, and Visha the poison."

"You forgot Uchhaisravas the white horse, Airavata the white elephant, Sankha the conch, and Surabhi the cow of plenty."

"How the fuck do you know all that?"

"I can read too."

"Anyway, so you know the rest. The Asuras were churning at the head of the snake and are believed to have been debilitated by noxious fumes, which may have been nothing more than bad breath. Anyhow, the Devas got the amrita and like Asterix and his fellow Gauls, on drinking the druids' magic potion, became very strong, etc."

"After which the good guys beat the pants off the bad guys, and retired and got pot bellies and became bureaucrats."

"Right."

"And Big Brother was happy because his man got to be president."

"Right."

### *The Lost Years*

*. . . I have felt them in my belly, churned them with my acidity, let them float in my currents, have felt the millions come and go, but I flow on, placid as the stars, unemotional as the universe, lap at shores, beat ancient rhythms against the rock, timeless tunes against silken shores, while my blood mingles with the millions within, I feel the hatchlings too, beneath my pores, under my skin, as they swim and swim in random confusion, for years and years . . .*

Jo passed the biscuits around and, tossing sticky sand out of her wild hair, leaned first against a wooden post and, finding it inadequate, against my shoulder. I held a hatchling up against the light, and we watched it flail its flipper. No larger than Jo's palm, it stretched its neck forward and strained to be free, to fly through the water.

"You know, for years, when scientists had no idea what happened to juvenile turtles . . ."

"Yeah, they went out as hatchlings and then turned up the size of dinner plates."

"So they call it the lost year. Isn't that presumptuous, thinking that it's lost to the turtle just because the turtles were lost to them?"

"History is the white man's history, you know—the dark continent, the dark ages, the lost years."

Turns out that it is actually years, and some turtles may take fifteen to thirty years to mature.

"Imagine hanging out all that time in seaweed rafts, FADS . . ."

"What?"

"Fish Aggregating Devices."

I sometimes think that we—Jo and I among others—were laid like turtle eggs. Dropped plop into the soft earth or someone's waiting hands and then abandoned. By the time we gain consciousness, we are swimming in unknown seas, buffeted by waves, carried by currents, and tossed by the tides. We hide in seaweed rafts and feed on the passing plankton, and catch in the passing wind some news of the world. In my dreams, I stare at clear blue skies from my raft. I have floated in this morass of weed and wood. A fad? A phase? Of the moon perhaps. It seems to me that I have been here for years, ever since I kicked off from my natal shore, and caught the waves square on my jaw, dived beneath the breakers, slipped passed the gaping jaws, squirmed through certain death, and then struck out for an unknown destination. But we were caught by a passing current, a blue moon tide, a swirl of determined water, a train of aqueous thought, a stream that knew its mind. And we were borne along, through hot water, cold water, warm fogs, clear skies, because it was easier than doing otherwise. Inertia is a comfortable coffin.

*These days I drift in and out of consciousness like a patient etherized upon the table by an anesthetist with a sick sense of humor. Every so often, I wake to the smoggy skies of a different city, the smoky sky of the same small apartment. In my shaky waking moments, I sometimes remember what it is that I am waiting for. I hear the sound of acoustic guitars, which is funny, because I play the clarinet. My mornings are misty and I can usually reconstruct them only from the dishes in the sink. Oh that's nice, I had a pizza for lunch. My lost years are a marijuana dream. The depression is not half-bad. It is the enthusiasm for change, the heightened hopes, the exaggerated excitement, and frenetic anticipation that something is about to begin that is almost unbearable. When I awake depressed, I lie curled up in bed, waiting to die or for the day to end, whichever comes first, and in the evening, I have a beer and a pizza and watch TV and let the day die on me. That's the easy part. But the mornings I awake possessed, to do this and that and the other, to move. The adrenalin requires its fix. And it is when I am rolling the weed that I am happiest—crushing it, feeling it break against my palm, and then shaking the seed out, spilling it on to the sensual wax paper, rolling and then administering final licks. The first drags and the final coherent thoughts.*

I did see Jo as soon as she was born, very briefly. Gina christened her Joanna; I decided to call her Jyotsna. That way we could both call her Jo. Not that it made a difference initially, since I didn't see her for the next eighteen-odd years, a period I spent alternating between frenetic activity and desperate inactivity. I had ended my short marriage to mathematics and was mindlessly working my way up a brain-dead corporate ladder, and I don't remember having a single coherent thought during that period. Jo spent a few incoherent years in the Midwest herself, an emotional wilderness born mainly of a distracted mother and an absent father. She came out to the University of California, San Francisco, the city of her conception, desperately searching for an intellectual and emotional home. Instead, she found me. Alternately, she may have come looking for me. She was a music student and had long fluttering fingers that played the clarinet exceedingly well. We met at a café, and she was glad to meet me, finally, etc., interested. She viewed that interview and subsequent meetings with a sense of detachment that disconcerted me.

"Do I have to call you Dad?"

"Amartya will do fine."

"The immortal one?"

I had told Gina that, and that her name meant "to live" in Hindi. Jyotsna was moonlight, what a little moonlight can do.

"And what do you call yourself?"

"Jo-Jo," she said and I laughed. She paused, then added, "though I sometimes think that yo-yo would be more appropriate."

"I'm sorry. Your mother and I didn't have the time to discuss the subtleties of naming you. Or raising you."
"She said that I was conceived at a guitar concert."
"Paco de Lucia, John McLaughlin, and Al di Meola, no less."
"At the concert?"
"Afterward technically. But the music was still kind of ringing in our ears."

We talked about her, how she had survived a series of small towns and ranches and grew up loathing cowboys and animals. I told her about my life—how I had suffered a series of jobs and cities and come to loathe nonresident Indians and software engineers, especially in combination. We shared marijuana dreams, a susceptibility she must have inherited from me. We bitched a bit about her mother. Nothing like a common enemy to get a relationship up and running.

I looked at the young girl with wild hair, all angles and accidents, pink too in parts, and knew that I had caught in the passing wind a smell of the past, a memory that beckoned.

*I dream that I am floating just below the surface, only inches below, all the time. The water ahead of me is murky but I know where I am. The weed is my home and as long as I can feel it around me, I feel safe and secure. The cloud is a blanket of comfort. Often I need to bob my head above the surface to breathe. My eyesight is poor, but I can see clear blue skies and stars at night and the sense of something stronger beneath the undercurrent. Something that calls from the earth's belly. Something below the breakers, below the tow, beneath the abyss. Something from the center. Of gravity.*

### Magnetic Memories

We lay under a sheet that was poor protection against a blistering wind that shot sand into our various orifices with no respite, and let our eyes roam the sky in which the stars brightened as the moon disappeared, searched for shapes, and created constellations. We were exhausted after counting and tagging several thousand turtles and were taking a short break.
"Where do so many fucking turtles come from?"

Kartik Shanker

"From far off."
"Very informative."

Green turtles in Brazil swim halfway across the ocean to nest on Ascension Island, which is a mere speck in the eye of the Atlantic. A spot of ink that god left on the globe by mistake when he was drawing the continents.
"But why do they swim all that distance?"
"Did you know that a sperm swimming up the vagina and into the uterus to fertilize the egg has been compared to a man swimming across the Atlantic in treacle. Turtles have it easy."
"No, really. I wonder why?" she said through her scarf-covered mouth, as she stared at the stars and generally contemplated great distances.

Beaches that are suitable for nesting are very different from ones that have food for turtles. One theory has it that when the continents were aligned differently, as one big chunk called Pangaea, the feeding and breeding grounds were much closer. But as the continents drifted apart, so did the beaches, and you know how particular turtles are about nesting on the same beach where they were born. The juveniles float around for a bit and eventually find their way back to the feeding grounds, where the adults forage during the nonbreeding season. The ridleys and leatherbacks are open ocean nomads and just wander around—until one day they are suddenly drawn home, magnetically inevitable, the males and the females. There is little courtship, no elaborate ceremony, no gifts, no flowers on the birthday. The male grabs the female, or in fact any passing object that grabs his fancy, and hooks his flippers around her, holding her with claws that are there for this purpose, bites her neck maybe, while his penis navigates the curve of her carapace, and finding her expectant cloaca, proceeds to complete the act of coitus. Orgasm? I don't think so. I have yet to hear a turtle roar or say *Oh baby.*

We eventually find our feeding grounds, and relatives, and memories of parents, and sometimes parents' memories.

Jo and I were seeking a common memory that would explain the years that we had lost. We decided to visit India, my family, the past she had not met, had not in fact conceived of. It would be, we were sure, an interesting experience at the very least. I was making a long-overdue journey home, playing the dutiful son, a guilt trip if you will. Jo was still seeking her niche, and she was a willing but dispassionate companion.

In Bombay, rather Mumbai, we stayed with my parents in their apartment in Malabar Hill, and spent the weekends at Marway beach, some distance outside the city, where we did not let the sea play games with our toes, for stronger men than I had disappeared into those hungry sands. But we did spend long evenings drinking vodka sunsets.

Jo was moodier than usual, and I felt I had to ask.
"What's the matter, Jo? Homesick?"
She tossed her head and said,
"Nope, you have to have a home before you can get homesick."
I guess I deserved that. And then she said,
"You know, Amartya, this beach is like any other beach in the world, just like the beach in California. We talk in a language that we both grew up with. So why do I feel so out of place here? So out of your life? Why does this mean something to you that it doesn't to me? Sometimes it feels like even the sea (and water at least must be the same everywhere) is accepting you and rejecting me."
We paused, the waves and I.
"And don't say it's all in my mind," she added.

We sneaked smokes on the terrace, looking at a complex Bombay skyline, typical of most Indian cities—one that mixes thatched patches of poverty with glassy facades of prosperity and is suitably confused. Not to mention dirty tenements where the streets are narrow, and drainwater collects in little gullies and floods during the monsoons. And the walls are sticky with graffiti and *paan* and painted stumps in front of which young heroes visualize winning cricket matches for their country, and the sound of the bat and the ball is background music. And yet we could not leave these places, did not feel secure enough elsewhere, were addicted to the smoke, drawn to the lights like moths or turtle hatchlings. Most people dared not move away, for fear of the unknown, the paradox of needing, in the same breath, the poison of carbon monoxide and the proximity of a medicine man who can do nothing about the damage that it causes. Jo and I were city kids, we loved Bombay and Calcutta almost as much as we loved New York and Chicago, but then we grew to love the beaches of Costa Rica and Papua New Guinea and Gahirmatha, hardly realizing where we were, except by the species of turtle.

*. . . So finally they have a mind of their own, even if it is nothing more than a magnetic memory, they start to swim against the currents, against the gentle persuasion that I push them with, against my irritated swells and my storming rages. Now they know where they are going. Ha. They think they know where they are going. When they reach what they remember as home, they will find that it is no longer there. I have shifted the sands from under their feet . . .*

We survived family and other animals at Bombay and Madras and a curious countryside in a small village in Orissa where my grandmother, the oldest member of my family, persisted, stubbornly defying modern medicine and ancient wisdom. We traveled by train, second class, the public class, trying to breathe the country in, the home that I thought I had left behind. But either it had moved or my memories were completely confused. We were no closer to finding whatever it was we had come looking for. We certainly did not believe that we had come all this way to find foul smells and dirty streets. The same smoggy skies that we had left behind on the other side of the Pacific. Or Atlantic, depending on the airline.

As I contemplated my roots in a dirty second-class compartment with the usual complement of bawling babies and sulky attendant mothers, it occurred to me that my grandmother must have traveled in some very similar landscape during the struggle for freedom. There were all kinds of struggles for all kinds of freedom in those days (as in these), and I am not sure exactly which one she had been involved in. But she could not have seen anything very different. The cities are different today, but these endless fields of paddy—the villages, the cows, dry fields, thin goatherds, a little grove, a patch of forest, a plot of land, and hot skies—these must surely have been the same.

*I wonder what the turtles saw, lying on their backs, hundreds, one on top of the other, as they were shipped by the eastern railways from the ports of Orissa to the pans of Calcutta under clear blue skies.*

Jo asked me if she could smoke, I said no, but she did anyway, standing at the door, and leaning nonchalantly while India and I looked on disapprovingly. Passing studs attempted to strike up conversation, the assumption being that a girl smoking in public is an easy lay. Jo blew smoke in their faces and sent them away with some atrocious German, a language she massacres, and told them that she was descended from Einstein or Goethe or some such persuasive Teutonic bloke.

We sat at the steps by the doorway and let the wind and the countryside slap at our faces. The shadow approached us obsequiously as the train turned toward the west and then slipped away suddenly, leaving us at the mercy of a mildly annoyed sun. Winter was just leaving this part of the country, and the sun was sharply pleasant for all of ten minutes. Jo turned red as a beet. But when the sun set beautifully over a rough tumble of scrub and thorn forest, some rocky hills in the background, I felt my nose tingle, and I knew it was the salt spray, a certain smell that the sea has, and I felt that we were headed in more or less the right direction.

### The Natal Beach

*Years later, when the moon or the tide or the wind whispers something, we swim against the currents, struggle against waves, rage against storms, and guided by our magnetic memories, find our way home. To a place we have never been before. To the beach where we were born.*

When we met Grandma, she was sitting in a long easy chair, her feet in the air, and her head in the clouds, hair perched on her head like a parrot about to take wing, in a large house in a small village, surrounded by paddy fields and the sea, a dark but airy house, with tall ceilings—and currently surrounded by indifferent-looking but insatiably curious neighbors. Grandma was still as sharp as a razor, though she kept asking me who Jo was, as if I might, if asked often enough, give her an answer that would satisfy her. Initially we spoke in brief sentences because our languages were so different, had become different. In the evening, I was waiting for the sun to bid adieu (so that I could sneak a drink in my room), when Grandma, her legs dangling still, called Jo to her side, held her hand and said,

"Ask him to show you the turtles."
And turning to me, with crystal eyes, she said,
"The turtles are dying, all of them. Your turtles. Do something."

On the beach the next day, the foul smell reached out and hit us between the eyes, sent shivers up our spines. The secret turtle burial grounds. Did they all migrate here to die en masse? Fatigue, labor pain, boredom? The beach was filled with putrefying carcasses, fresh ones, soggy from days in the sea, with bulging eyes and ballooned necks. Sundried carcasses, skeletons, a carpet of decaying flesh, a house of maggots, a dinner table for crows and dogs and vultures. Kilometer after kilometer, the turtles lay piled up. Some were buried; some had washed up fresh. But there were thousands.

*I had not thought death had undone so many.*

"My god," Jo said, gagging, "I didn't even know that there were so many turtles."
"Trawlers," Grandma told us when we went back. "They've taken all the fish, and now they're killing all the turtles. Incidental catch. They drown in trawler nets, and their bodies are thrown overboard."

Trawlers are supposed to use turtle excluder devices—TEDs—on their nets, trap doors that let only the turtles escape. But few of them bother; none of them cares. A few install TEDs for the sake of the public eye and the legal eagle, but later they tie the door with wire, to prevent even that minimal loss of fish.

"Bastards," she said vehemently. "Think of those turtles, so small when they swim out, the little ones, all those years. They roam around looking for their home, and finally they find it and come back to lay their eggs, only to drown in some fool's fishing net. All those years. All for nothing."

She talked to us about turtles for three hours—turtles and crocodiles and people who were some of both—people who had left and lost, been lost, people who had not returned. Only one in a thousand survives. Grow like trees, she said. Let your roots grow deeper even as you reach for the stars. Fly and be free, I told her, like birds or turtles. She shrugged, acknowledging this world-view, but said, even turtles will come home to nest. Jo sat by her side and held her hand and looked happy but uncertain, confused but calm. I helped her back to bed, still cursing under her breath, so small and frail, lonely as a turtle, and Jo, feeling small and frail, tucked her in, and we left her flailing her flippers at an indifferent world.

*Flippers are a hindrance on land. They make you awkward, clumsy, ungainly. Vulnerable. In the water, they make you an artist. A ballerina, you can pirouette, and spin and dance through the water, sleeping beauty and swanlake your way through the thick blue stage. You can hang and glide and soar and dive, and then surface with a burst of joy into the arms of the waiting sun, gulping air, before slipping like an elusive love, into the mysterious medium below,*

Late in the night, full of anger and grief for turtles and other creatures with lost years, Jo cried in my arms. All this while she had been detached and distant, an indifferent observer of her past, had let me light her cigarettes when she wanted to indicate affection. But tonight she put her head on my shoulder and cried the tears not shed before on these shores, for shoulders not leaned on, for bums not wiped, for the fights and the conversations and years that had not been. And when I pulled her close to me, she pressed her cheek to mine, and I cradled her head, and felt her warm and soft under her thin cotton dress. I did not let myself wake, and neither did she, for this was something beneath the conscious, below the tow, something from the center. Words are weary, thought is scary, imagination runs headless to the precipice and stands teetering on the edge. Jo looked up at me, her eyes twinkling *Noctiluca* that I recognized from somewhere, and a current ran through our barely insulated bodies. And then we were underwater, sinking and swimming in a swirl of the blue moon tide, and the sunlight filtered through from above and lit up the corals around us. It was a time in our evolution when we had not left the water, when we were fish, or maybe turtles, a time when it did not matter who or what we were and only the moment did, when you burst with joy into the arms of the waiting sun. I remember her soft young lips, pliant and plaintive, seeking a safe haven, and much else that was soft and searching for refuge in a random universe. And later, there were turtle tears on our cheeks.

*I have been swimming for many hours, it is time for a pause, a bask in the sun, a back-baking exercise to recharge the batteries, when suddenly . . . I am surrounded by a thick mesh, a net that seems to move past me at an alarming rate. Initially I am unafraid, the water is after all my home. I swim up, but the net reaches above me, and below, and left and right. It stretches for miles in front and behind, and I see no way out. But now there is a light at the end of the funnel, a metallic glint, a last hope. As it approaches, I am shorter and shorter of breath, the water feels heavy around me, like treacle. I reach the end and see the door above me, my salvation, the only way out and I swim up against it, but it will not move, I thrash wildly against the door, the net, the fish, there are other turtles here, thrashing too, some are lying heavily against the bottom of the net, and I slowly sink down beside them, and as I begin to slip away . . .*

I dream of blue skies and starry nights, I have a memory of a place, a place I have seen once before. Once enough to know that it is where I have been heading all these years. All the time, hidden in the weed, I had not remembered. But now, the sky grows bluer and the stars light up every single point in the sky. Until there is nothing else.

I wake up sweating, breathing heavily and gasping. Reach for the water and drink slowly. Jo turns in her bed, and I slip out quietly.

The compass in my head has brought me here, the magnet in my brain has guided me, but suddenly as I see the dunes in the distance, I smell a memory, which is vivid and distinct—the smell of the soil that I climbed through, the beach that I scrambled on, the water that I rushed headlong into.

It is a moonwashed night, and I am a little awed by the silence of the world around me. We stand at the water's edge and let the sea push sand between our toes. I clutch her hand, the nice lady with the wispy white hair who has been sneaking sweets to me all day, chewy brown balls, milky soft delights, and my hands are still sticky with the memory of the day's excesses. For now I am glad to let her lead me to the water. In the distance, the fishermen are getting ready to go out to sea and begin the day's work. They drag their catamarans out calling to Grandma as they heave their ropes and nets and sail, and push the boat out to sea. We wait until the light dies, and then Grandma says it is time to go home, but we stand rooted to the sand, our toes magnetically bound to the earth there. A wave finally breaks the spell, and we move away quickly to our brighter horizon.

Sticky feet, sand under my nails, we shuffle back homeward, an old lady and a little boy. Suddenly she stops and points. In the moonlight, in front of me, is a large unmoving mass, a rock. We move closer slowly, and she squats on the sand beside the object, holding me close. It is a turtle, she says, a mother turtle laying her eggs in the sand. I watch, unable to take my eyes off as she periodically lifts her head to take deep gasping breaths, and salt tears trickle out of her eyes.

We watched for a while, and then Grandma called out to a fisherman, who dug a hole behind the turtle and put in his hand and pulled out a few eggs and held them out for me. Soft and wet with mucus. I held my hand under her, and the eggs fell, soft wet and stirring, into my minuscule hands. The fisherman wanted to take a few eggs away for dinner, but Grandma sent him away with a flea in his ear. We watched as the turtle covered her nest, and I followed her as she waddled back into the sea, into the water, until Grandma grabbed my disappearing fingers. Would have followed her much further if . . .

And now, climbing on to the beach again, I realize that this is where I started all those years ago. I have found my natal beach. I do not know how many times I will be able to come back here, how many times I will escape the murderous nets and the vicissitudes of an amoral ocean, but once is enough for my soul.

I sit cross-legged beside the nesting turtle, feel her dropping eggs into the sand below. We breathe and sigh and exhale slowly, salty tears in our eyes. When she is finished, I walk her back to the water. We stop to breathe and prepare ourselves and then move smartly on toward our brighter horizon. She kicks off from uncomfortable shores into her ballroom sea, and I stand there at the edge playing footsie with the sea and the sand.

### *Coda*

She turned her head to me, and I felt her forehead against my chin, felt a prickle at the back of my neck, a little tingle that ran up my spine and tickled the hair at the nape of my neck. She reached for my hand and interlocked her fingers with mine. I felt a pang as I realized that I was as powerless as a turtle to protect my hatchling, but I moved my arm around her and held her close.

*No matter how large you grow, how mature you are, the sea is always infinitely larger, and frankly that scares the shit out of me. I am always running for the seaweed raft, but sometimes the haven is not what it seems. On the other hand, the big green monster who lurks behind the shipwreck is not always evil. At least I know that home is not a figment of my imagination, a metaphysical concept, a legend that people created, a faith that they constructed, like God, to make life easier. It may not be easy to find, but it's there, even if it's only a magnetic memory that I will chase all my life.*

# The Limit of Love

*David Morse*

Max and Esther arrived at Delphi by bus, twenty-five-hundred years too late. The oracle was long gone. The crack in the earth from which hallucinogenic vapors once wafted had closed without a trace. The sacred springs were dry. And for that matter, after the long ride from Athens, the questions in Max's mind required no divination. Would the place be mobbed? Would his bowels move in timely fashion? Would Esther remember to take her eyedrops?

When their bus pulled up in front of the Sanctuary of Apollo, Max was disappointed to see half a dozen other buses already disgorging passengers—mostly Europeans, to judge from the set of their faces—all of them doubtless as eager as he was to beat the crowd. This is what's meant, he thought, by the tourist industry; this pipeline of interchangeable components saying gee whiz in their various languages.

Nina, their tour guide, stood up from her seat at the front of the bus. Her microphone crackled as she smacked her lips and huffed. She was a short, stolid woman with halting English and that distracting habit of lip smacking as she chose her next phrase. Earlier, as they crossed the Thebian plain, she had related the story of Oedipus so blandly that when she got to the part where Oedipus, smack, "took out his eyes," he could have been taking out his credit card. On the other hand, she had referred to the bombing in Kosovo with scathing irony. Bombing (she pronounced the second *b*) the people to save them. Now she announced that she was returning to Athens, so they would be fending for themselves. She put it more euphemistically than that. "You will have free time."

"Nothing oracular about that," Esther observed wryly.

"Free time we paid for," Max grumbled.

They were at a grumbling stage of life, or at least of this tour. They stopped to use the toilets, filled their plastic flasks at the drinking fountain. Max recalled how in the fourth grade at recess, Mrs. Burdette used to tell them to tank in when they were lined up to use the drinking fountain, and once when they were lined up at the restrooms she told them to tank out, and laughed at her own joke with the other fourth-grade teacher.

They followed the Sacred Way uphill past the various marble ruins, hoping to attach themselves to a tour being conducted in English, but the babble around them was in German, French, and Greek. Esther understood enough French to translate for Max, but he grew impatient with the process, knowing it was a distraction for Esther, annoyed at himself for losing his French, even more annoyed at the loss of hearing, which caused voices in a crowd like this to well up around him, shrill and metallic, as if he were standing inside an enormous saucepan. He hated growing old. This morning, looking in the hotel mirror, he realized he was tired of his own face. The skin that trailed over the fold of his eyelid, the very familiarity of it, the secret fatigue in his gaze, caused a kind of shudder in him. Was this what it was to grow old? To feel increasingly out of sorts, outside oneself? The voices rattled around him. Bright shirts, cigarettes. The contrast of sun and deep shadows baffled his gaze.

At the stone amphitheater, he told Esther to proceed without him. The amphitheater looked peaceful. "I'll meet you up at the top, at the stadium," he told her, pointing to the map of the site.

"Are you sure you'll be all right, Max?" She cocked her head, appraising him.

"Yes, yes." He'd had a minor episode in Athens, a moment of befuddlement and dizziness, walking near Syntagma Square, where a crowd of demonstrators was chanting. A banner showed the silhouette of a bomb with the word NATO on it. Dizzy, but too embarrassed to sit down on the sidewalk, Max had clutched a signpost and shut his eyes until the dizziness passed. Are you all right, Max? Esther's voice strange in his ear, the world wanting to turn upside down around him. Only the signpost kept him upright. And then the moment passed, and he recovered completely. Dehydration, they decided. After that they had made a point of carrying the plastic bottles of water everywhere they went.

Esther's pale blue eyes regarded him seriously. "Drink water," she said. "Keep your hat on." She was wearing her yellow sunshade.

He nodded, tapped the brim of his NPR baseball cap, thumped the plastic flask. Saluted.

She smiled impishly and turned to leave. He was glad to be left alone. Yet he felt a qualm, watching Esther hurry up the hill behind the French group—a moment of fear. Quite unexpected, even exotic. They had separated on other occasions. He had walked through the Plaka by himself while she napped in their hotel room. But this was the first time since that odd business at Syntagma Square. What was it? Fear of dying, alone, surrounded by unfamiliar faces? An irrational thought. He smiled at himself and took a deep breath, relieved to be free of all those voices. It occurred to him that he and Esther had not arranged a backup time and place, as they usually did. But there seemed to be only the one path leading up the slope, and her yellow sunshade was easy to spot.

Behind him on the Sacred Way, a group of Germans was assembling by the Temple of Apollo. Soon they would overtake him. All precise questions and guttural earnestness. But for the moment he was able to appropriate the quietude of the stone amphitheater. He stood on the circular stage, which was perhaps twenty-five feet in diameter, and looked around at the tiered benches rising in a semicircle, the blue mountains swelling beyond.

He placed himself in the exact center of the stage, the spot where lines extended inward from the aisles would have met like cross-hairs. The place was not perfectly symmetrical, he noticed. He liked that. Mount Parnassus loomed steeply behind him, its flanks swelling more gently on either side, cradling the amphitheater in its broad, uneven lap. The tawny slopes were peppered with wildflowers. Poppies scattered everywhere like the red peckings of a marker-wielding grandchild and daisies and small yellow flowers of all sizes. Goats grazing in the distance. What a magnificent setting! Even now, arid as it was, deforested for olive oil, he could feel in the very balls of his feet why the ancients had chosen this site. The enormity of standing here at this intersection of invisible lines swept over him. This was the font of Western civilization. Three thousand years of written history stretched behind him. What would it be like in another three millennia?

Below was the ravine that was once a stream, fed by the great Kastalian Spring, flowing into the Gulf of Corinth, through which the ancient ships arrived from across the Mediterranean as far away as north Africa—all converging here, on this spot: the navel of this watery world. The Kastalian Spring had ceased to flow once the olive forests were cut down. The river harbor where ancient ships had once moored, probably within sight of waterfalls, was now a ravine and nothing more. Everything was dry.

Max clapped his palms together smartly, hoping to hear an echo, but there was none.

A woman gazed at him from the shadow of a cypress, her white hat and blouse glowing in the cool light. How long had she been watching him? She stepped from the shadows. Perhaps she was waiting her turn in the circle.

He beckoned her closer. "Do you speak English?" he asked.

"A little," she said. "You are having a—what do you say? Spiritual experience."

Max felt his ears draw back. He would not have used that word without irony. But he could not divine her tone, nor could he identify her accent. "Yes. I suppose." He chuckled.

She was young—younger than his daughter Becky, who was forty-two. Under the shade of her broad-brimmed hat, her skin looked Mediterranean, dark hair pulled straight back in a ponytail. She wore jeans, a loose linen blouse. She was tall, her carriage graceful, the clarity of her gaze striking.

She too had observed the asymmetry. "The builders respected—," she searched for words, moved her hands, palm down as if weighing something. "—the balance of the land, eh?" With the expansiveness of a dancer, she turned to indicate the hills and the valley.

He nodded. "A sacred place." Not a word he would have chosen normally. Something about the woman elicited it. And yet, was that not what he had been feeling?

"It is where the tragedy was born."

He wondered if she was referring to Nietzsche, the origins of Greek drama. But the Germans were invading, a noisy tide rising almost on cue. Max and the woman traded wry looks. She glanced up the path in a way that invited him to accompany her. Wordlessly, they proceeded uphill. Max's steps quickened with excitement, pleasure at finding himself in this young woman's company.

They located the ancient spring of Kerna, a smaller spring, also dry, with niches cut in the stone apparently for offerings. They took turns peering into the cavity, damp with moss. Something erotic about this, Max thought. How did she experience it? Aloud, he wondered what minor deities—water gods, river nymphs, dryads—would have been associated with this spring. They consulted their guidebooks and found little information.

"I think to purify," she said. She pantomimed washing her arms and face. Looked up and smiled. He pretended also to wash. They performed these ablutions in silence. If Esther could see them! Chagrin at the thought. Delight. He had the opportunity to observe her eyes more closely. The irises were gray-green, striated, the whites exceptionally clear. Flashing-eyed Athena. Her steps ahead of him, as they continued up the path, were athletic and yet leisurely. She stopped often. Was she allowing him to rest? The sun was high. Perspiration prickled his forehead under the baseball cap. He offered her a drink of water from the flask, took a swallow himself.

He coaxed her to talk. It came naturally to him; all those graduate students he had taught. She was here at Delphi with her friend. "My fiancé," she corrected herself. "He climbs the mountain—"

"Parnassus?" But it was the other word that Max registered—fiancé—surprised to find himself somehow disappointed. Abashed, he chuckled to himself. No fool like an old fool.

"Yes, yes. Parnassus."

He found himself looking for Esther's yellow sunshade. Wondered: Has she reached the stadium? Will she see me with this young woman? Not all those lunches with graduate students were entirely innocent. The eyelash in the eye. The invented committee meetings. The letter Esther found on his desk. But this woman, younger than Becky! He laughed at himself, at his fantasies. He did not feel old.

And why was she not climbing Parnassus? Was she quarreling with her fiancé? "Why—," he started to ask, but stopped himself. It would be prying. And perhaps he did not want to know. Preferred the fantasies. He looked up at the gentle rolling flank of the mountain and thought of her lover climbing alone. Max felt a comfort about the place, a familiarity—as though he had been here before, played in fields like these, fragrant with blossoms. Was his memory so easily fooled? Recovered memories. Or was the world so small?

"Why do I not climb with him? This is what you want to know, eh?"

Max nodded.

"I am ill. I was ill?" She frowned, struggling with the tenses, dissatisfied. "I am conva—How do you say? Conval—"

"Convalescing?"

"Just so." She sighed hard, then said the words: cervical cancer, radiation. Max took the words in cautiously. She was so young! Younger than his daughter. And with a fiancé.

"One cannot know!" she concluded, tipping her chin up in a way that made him think she was Greek. Why was she telling him all this, a stranger? Was she that much at home with the terror? He was almost annoyed. "No," he agreed. "One can never know."

"It is to wait, and hope, eh? One can only live. In any case, I am here at Delphi! Now!" She swung her arm around to gather in the vista of hills and ravine that led to the Gulf of Corinth. "So we climb the different mountain. He climbs his Parnassus, and I climb mine, where I can ask my question!"

"What is your question?" As much as he and Esther had joked about it, offering to bring their friends' questions to the oracle, Max felt the lack of a question now as an omission, a flaw in their characters.

"It is this." She smiled, standing with her feet spread. "What is the limit, eh?—the limitation—of love?"

They stepped aside to let a tour group pass, while he supposed he should be pondering her question, the din of voices scrambling his thoughts. He avoided eye contact with her. If she spoke, he would have to ask her to repeat herself. He had difficulty holding the question in mind. What did it mean—*What is the limit of love?*

All the great questions his students used to ask. The undergraduates, especially. Why do we live? Is there a consciousness that transcends our own? All those questions that adults gradually abandon as unanswerable, or answer by some crutch of faith, or by some process more amorphous than faith give themselves permission to let fall into the compost of younger minds. Those questions, he reminded himself, are still alive for the young. They are the questions that drive the species at some grand level, the questions that drove Delphi.

He used to suppose that a few of those questions would get answered with age, that knowledge would accrue like the growth rings on an oak. But so far he had discovered that it was not the answers that arrived over time; it was an evolving understanding of the way the questions got asked, of what they signified in the life of the asker. They meant nothing outside a context. And it was the little questions and the little answers, finally, that bore everything forward: Shall I put milk in my coffee, or cream? Can I get away with wearing this undershirt another day? How fast is that car coming? Those were what mattered finally, ridiculously.

Max felt the gulf of years between himself and the young woman shimmering in the space between them, suddenly more palpable than his own weariness in the mirror this morning, which seemed a distant apparition. Would he choose, if he could, to be as young as she? He thought not. He would go back ten years if he could, but not forty. One lifetime was enough.

They stopped at a railing and looked back down the way they had come. From this elevation they were looking down at the amphitheater; beyond was the rectangular foundation of the Temple of Apollo with its broken marble columns; below that, various treasuries, all shining in the sun; farther down, across the two-lane asphalt road, lay the smaller and more ancient sanctuary dedicated to Athena Pronaia, its double row of pediments perched on the edge of the ravine; and finally the glitter of the Gulf of Corinth.

"What was it really like?" she mused. "In ancient times."

Max removed his baseball cap and mopped his forehead with a bandanna. "Mobbed, I expect, on holidays."

"Like now."

"Worse. Think of it! All those athletes and noblemen with their retinues arriving by the boatload from all over the Amphictyonic League—from Sicily and Libya and Turkey. Imagine the babble, many of them unable to understand each other. Bringing precious offerings and animals to sacrifice." Returning his cap to his head, he gripped the iron railing and shut his eyes. "Think of the clutter in this confined place. All those statues. All of them brightly painted. I hate the idea of the ancient Greeks painting their statues, putting jewels in the eyes. Everything was overdone."

"Overdone?"

"Vulgar excess. Over the top. All that statuary jammed together, some of the pieces gilded—every size and shape, commemorating this and that god, and who knows what politics behind the theology. Pretentious, gaudy stuff designed to impress. Griffins atop columns. Bronze tripods and incense pots." He was lost for a moment in that swirl of color and sound, felt a trembling in his knees, an unsteadiness as he gripped the railing. He was aware of his own voice, incantatory as the water he described spilling from one marble pool to the next. Tons of sacrifices, carcasses of bulls heaped up on the stoa. He could smell their guts steaming. Priests examining the entrails of birds. Blood running in the gutters. All of it running down to the sea. A great stink of death. He could hear it in his voice. Smell it.

He listened, detached, to the stream of his words. "Down there at the Temple of Apollo you can see the priests are out on the steps, weighing gold. Warriors parading about in their finest armor. Slaves carrying rolled tents, amphora, carved chests. Look at the baths. Peddlers squatting by blankets with charcoal braziers, hawking grilled octopus and souvenirs, sacred springwater! Laurel sprigs tied up with orange thread! Lamb shishkabob!" He laughed. "There's your sacred place!"

The vividness of it faded at the sound of his laughter. He opened his eyes, relaxed his grip on the rail. Thought of the crowd of protestors at Syntagma Square.

She looked rapt. "That is a beautiful vision."

"Beautiful? Hmmph!" Max shook his head. "I like it better as ruins. The bleached marble torsos. The empty eyes."

"Ah, but your spirit is in the other!"

"Some part of me perhaps."

She was silent for a time. "I take you my question."

"What question?"

"What is the limit of love?"

He chuckled.

"I think you are the oracle." Her eyes flashed. "So. You I ask my question."

He saw that she was serious. "You must explain what it means."

She nodded. "Something like this, if you forgive my English. I believe that when you confront death, you confront life. You see how we are alone, eh? You feel the limits of love. And you also feel the strength of it, because you are—how do you say?—engaged with those limits." She laced her fingers together and shook them. "You are living in that engagement. In that edge of the self. But we are alone. Do you understand me?"

He nodded, thinking of the fiancé climbing somewhere far above them. He understood she was employing the word *engagement* differently. Still, Max wondered. Had they become engaged before or after the cancer? And he was remembering when Esther's sister, Fanny, died, how he and Esther had lain together in bed, hugging—they who, for all their arthritis and glaucoma and loss of hearing, had neither one experienced a life-threatening illness—hugging as if their bodies could somehow fuse. And still they were alone.

"So." She was watching his face. "What is the limit of love?"

Max's hands felt empty, hanging at his side. An old man's hands, the skin crinkled like a chicken's. He sighed. "I don't know. I have no answer."

She looked away, but then returned her gaze and nodded. "Perhaps it is the answer I need. Perhaps another answer is not so honest."

Max was indeed trying to think of something less honest. Something oracular.

She smiled. "In any case, thank you. You give me something."

Max felt his ears get warm. What did she want? "I give you something?" He repeated it as a question, confused. Could she be asking for money?

"Yes. Right now. You give me something I never forget. When I think of Delphi, it will fill my mind as you tell it—so, how do you say, over the top?" She waved an arm at the scene below and her whole face beamed, and he knew it was true.

She turned to leave, but first she took his hand and pressed it. Before he could think what to say, she disappeared into the crowd. His hand tingled with the life of her touch.

Arriving at the stadium, he scanned the long row of stone bleachers overlooking the playing field where half a dozen young men were showing off, running footraces. Among the onlookers he spotted Esther's yellow sunshade. One of the runners had removed his shirt; eyes squinting from sweat, chest gleaming in the sun.

*Hot Water Series.* © Margaret McCarthy

# The Sea Changes

*Christopher Woods*

Here, this night, Alexander's letters rest on my desk. They are scattered here and there, in small piles around the room, on sills and shelves and the floor. When I walk, I feel I am walking across his heart.

But he won't mind. After all, Alexander does not expect me to remain seated at my desk, reading and rereading his letters, as if there were nothing else to do. He would not want to limit my life in that way.

Still, there is no getting around it, how much time I spend on his letters. They are so pure, from deep inside him. Some I have memorized. All I have taken to heart. A few, the most private of all, are curled in the pockets of old pants. I carry them with me on long walks. When I sleep in old pants, Alexander's letters are with me to walk through my dreams. In this way, they are like directives.

All these letters (who can know how many?) are growing older. Like myself, like Alexander. The yellowed envelopes are pressed inside another, even larger envelope made of time and dust, of old light and dimming dreams. Looking at them, and at all that is now my life, some nights I imagine the letters to be layers of Alexander himself.

The first layer must always be memory. It dances and hides and is the hardest to know for certain. The next layer must be time, though these days that is but another uncertainty that runs a course, maybe oblivious to me, perhaps in spite of me.

No doubt there are other layers, many of them, but they are not so important here and now. Memory and time, they must matter most. I put the one with the other, make a ball of them, then roll it from one palm to the other. This is how I try to make sense of Alexander and why he matters so much to me.

This is a hard thing to know. So much time has drifted by and away. Time, friend, is river water. Time is wind. It is a force that prevents things from staying as they are for very long. Time is Alexander, or he is time. I suspect the truth lies somewhere between the two.

I like to think of him still young and unchanged, like the sea mountains where he lives. I close my eyes and imagine him there, in a small house nestled between blue sky and low, scraping clouds. He is gathered in the hard hands and arms of those mountains.

I think of him at night, perhaps frightened in that silent place of wavy, liquid wind that rushes through crag and bone. I like to think his letters are born of this same silence, and perhaps fear.

I wonder why he is condemned to live in a place most would call dire and forlorn. Here, so far away from all that, I wait to see his postmark, a gnarled oak root strangling a snake. His letters are strikes against the silence. They are a way to confront evil, to defeat it.

Each letter, when it does arrive, shatters the silence of my own sea room. Through the afternoon I watch the sun, crimson with longing, as it plunges into the cool, wide bay. I can't change any of it. I say nothing, not to people in the village, not even to myself. I have, in fact, spoken to no one in years.

This is because, a long time ago now, I rode that red sun into the bay. Many say I was lost, that I never returned, that I retreated in the spiral world of the conch. All I know is that it was time to retreat, to turn inward.

Yes, I am alone here. If I am in some kind of hereafter, so be it. I have no qualms about it. What matters most now are the letters from Alexander. I shift in my chair in the cold night, the dark watery expanse all around me. I feel the waves carrying me so sensuously, from one palm to the other.

I continue to write letters to Alexander, and hope that he awaits them with the same longing as I do his. I try to imagine the delighted look on his face when he receives them, how he holds the pages in his hands.

When I write, I print well, taking great care. When a letter is complete, I add my seal, the gnarled oak root strangling a snake. Tomorrow, I tell myself, I will mail it, and I truly believe this will happen. I place the letter on one of many piles that only grow, on sills and shelves, on the floor.

*Rama Cay, Nicaragua.* © Helen M. Ellis

## Swimming with Children and Three Hundred Dolphins

*Anne Collet*

*Translated by Gayle Wurst*

When I'm at sea, I like to keep the night watch. And on the *Fleur de Lampaul,* I like to keep the night watch with the children. I am amused by the sleepy expressions on their faces as they put on their slickers and equipment before going up to the deck, by the way they pull their hats down all the way over their eyebrows. Nighttime impresses them—little inclined to talk, they cock their ears in the attempt to hear the wind welling up in the darkness. Often the whales are there—we are, after all, on their "territory"—but most of the time there is only the murmur of water against the hull, and the whisper of wind in the sails.

Still heavy with sleep, the children sit on the damp wood, knees tucked under their chins, hands wrapped round their ankles. They shiver and swallow hot chocolate in short little sips, their gazes lost in the stars, while the *Fleur,* her sails shortened, travels around in a circle. By remaining in the same area, tomorrow we'll find sperm whales and dolphins again. All around us the moonlight illuminates a misty, shifting universe. The sea at night is not for humans, unless it is viewed from the deck of a ship where nocturnal terrors can be defied in relative security. But this unique time weaves strong and invisible ties, creating a sense of complicity which will prove to be a precious asset when the moment comes to dive among the animals at dawn.

Some of the children sleep on the sly, others come over to me asking in whispers if they are really going to see dolphins, if they are really going to get in the water with them, if it will be hard, if it will last a long time, if it won't be scary to swim with them. We have hours before us with nothing to do but talk, with long periods of silence to give the words time enough to grow and nourish their dreams. In the darkness, their voices move me. As I listen to them, I imagine their desires and fears. They remind me that I, too, once was ten yours old, and that I dreamed, like them, but beneath other stars. It was a time when Cousteau's camera had just transformed the horrific abyss to a wonderland, when no one yet imagined that common mortals could pass through to the other side of the mirror. Back then, thirty years ago, factory ships in the Southern seas were killing whales by the thousands, and porpoises snitching sardines from fishing nets were hunted down with rifles off the coast of France. Unlike cats and dogs, horses and a few birds, cetaceans had not yet been admitted to the pantheon of animals thought to be "man's friend." We were not yet capturing orcas and bottlenose dolphins for aquatic circuses, we were not yet driving them to neurosis for the sake of making a splash with crowds of curious onlookers.

When the future adventure seekers on the *Fleur de Lampaul* first came to meet me at the Center [Center for Research on Marine Mammals] I already felt attracted to the project Charles Hervé-Gruyer, the owner of the *Fleur*, had conceived. He was ready to embark a dozen young sailor-reporters between the ages of twelve and fifteen on a voyage to discover the planet. This project included exploring the world of cetaceans, a subject which was very apt to interest me. But I do not, a priori, feel the calling to incite my peers, much less children, to approach marine mammals physically. Too many experiments with the goal of establishing a so-called contact between humans and dolphins have had disastrous effects, especially for the animals. The same sometimes holds true for the mental health of those who participate in the experiments.

Nevertheless, in our very first discussions at the Center, the team that Charles Hervé-Gruyer had put together began to calm my fears. I talked with the young people, I described the skeletons of dolphins we had carefully reconstituted from washed up specimens, I made them take a tour of the organ samples we keep in our laboratory, I spoke to them at length about what it was like to work under the microscope, cooped up in a narrow room, or to stand on the deck of a ship staring at an empty sea through a pair of binoculars. I foregrounded all the worst aspects of a scientist's work, and yet their response was unexpectedly sympathetic. They were supposed to come for an hour or two, but every time they visited the Center, they stayed the whole day long. They questioned me, took notes and photographs, and learned as much as they could, accumulating knowledge with a view to future encounters with dolphins or whales. I saw that their organizers were just as serious as they were. They had undertaken a project in the spirit of true pedagogy, imagining an intelligent approach to cetaceans in their natural habitat. I invested a bit of my own time, and was amply repaid by the curiosity shining in the children's eyes and their ever-pertinent questions.

Later, when they asked me to accompany them on their expeditions, I hesitated from lack of time, even though I had always considered it my duty to leave the lab as often as possible to explain the results of our research to newcomers. To my way of thinking, such availability is far too rare among scientists. In the end, the children's motivation, passion and thirst for knowledge convinced me to help them "pull some strings" to arrange a meeting with the whales.

Later still, like every other passionate sailor who ever set foot aboard that wonderful ship, I fell in love with the *Fleur de Lampaul*. It is a national treasure.

The expedition to the Azores in 1991 took place during summer vacation. There was no need for the children to pursue their studies on board, which left us more time to learn about marine animals in the best of conditions. In addition to its beauty, the *Fleur* is an excellent floating observatory. A nineteenth-century sailing ship once used to transport sand, it is robust, spacious, quiet, stable, and high on the water. The friendly yet responsible atmosphere on board incited us all to do whatever we could to provide our youthful adventurers with what they had come looking for—as long as the animals also wanted to cooperate.

The back of the mid-Atlantic mountain chain surges up from the depths of the ocean, breaking the surface 1,500 kilometers from Lisbon. The nine Portuguese islands which constitute the Azores are among its highest summits. Cold currents originating from the bottom cross warm ones from a branch of the Gulf Stream in the vicinity of these volcanic islands. The water is temperate and the weather is capricious even in summer, but this is one area on the planet where cetaceans are sure to be seen, swimming in waters that the surrounding islands shelter from the heaviest swells. The sperm whales are there, of course: it is they who created the glory of the islands—Pico, Faïal and Saõ

Miguel, where generations of harpooners known for their incredible recklessness followed one after the other, up until the time that whaling was prohibited. Bottlenose dolphins are also present in great number, along with common dolphins, Risso's dolphins, short-finned pilot whales, and above all, Atlantic spotted dolphins. More rarely, it is possible to see rorquals, killer whales (orcas), false killer whales, and Sowerby's beaked whales.

The spectacle of mountains plunging straight into the sea and rising up into the mist is a source of constant pleasure in the Azores; but it is simply out of the question to anchor near their shores, for the water is far too deep. If we were to be ready to go to work at dawn, the *Fleur* couldn't leave the area where the animals would appear. According to the winds and currents, we let her gently drift or tacked throughout the night. Although I am unable to offer any rigorous explanations, our best contacts with cetaceans take place immediately after sunrise. At dawn, when the sea is still, the animals are apparently sensitive to the peacefulness, and seem calmer themselves. They accept our boats, whose motors they hear from afar, with no sign of surprise, nor are they surprised by our presence. We do not understand why they are most convivial in the hours just after dawn, because we don't yet know the rhythms by which different species live. Certain animals that only feed at night may experience a period of rest in the morning, when they are freer for encounters or games. But others, known to hunt both day and night, are just as peaceful, and just as curious about us. For whatever reason, dolphins most often come calling at daybreak.

But they are also quite capable of acting aloof. Dolphins refuse to fit the categories which a literature created for "dolphin lovers" often tries to impose on them. Pompous systems and mystical claims are reductive in any case. Dolphins are highly evolved mammals whose intelligence has permitted them to go beyond the droning uniformity of bees or ants, as amazing as these creatures may be in their own right. From one hive to another, from one anthill to another, insects of the same species all act pretty much the same. But not dolphins. They are like those grammars whose every rule is accompanied by a long list of exceptions. All attempts to classify them according to iron-clad laws of behavior have failed. Every study shows that whatever we learn from one group of dolphins does not apply to the next. And all the wild imaginings about dolphins being almost human—or more than human, close to divine—are equally without a future. While dolphins are considered as social animals which live in families and groups, others remain solitary for long periods of time, sometimes for several years, living in the same territory. Many scientists consider them as fortuitous cases, and thus negligible. Others, gullible fools or sorry profiteers of cheap mythology, sometimes call them "ambassadors," as if they carried a message from their "people" to "ours," or played an intermediary role between the human and the divine.

Fortunately, dolphins themselves will never know the heavy freight of commerce they carry on their backs, or the fantasies with which we've adorned them. Such excesses and blatherings only go to show that our knowledge is still very limited, that in certain domains we haven't even begun, and that ignorance has always been the most fertile ground for illusion. When science cannot explain, the field is wide open to imagination.

The more our research advances, the more we discover that the behavior of species, stocks, pods, and even individuals differs from one to the other. We are still very far from being able to formulate rules, if rules indeed are what we want. For the moment, we are happy to discover more and more new possibilities, abilities, strategies, and techniques that a dolphin uses to live in its own way, like an artist who creates an infinite variety of hues from a painter's palette. The behavior of orcas, for example, among the most popular cetaceans all over the world, still remains a mystery to us. Some pods master skills of which others remain completely ignorant. Certain of them feed on fish, while others eat seals. Within any one group, some individuals are more gifted than others in approaching their prey, camouflaging themselves, or even voluntarily beaching themselves to capture a young elephant seal or penguin. Bottlenose dolphins found in Florida live differently from those along the English coast, while those along the shores of Brittany behave otherwise than the ones found near Arcachon in southern France. Starting in 1996, we began to study dolphins in the Charente straits because chances are that their behavior demonstrates yet a different strategy. And in fact, preliminary results do not conform to those obtained with dolphins near the neighboring Isle de Sein or in the Archachon basin.

Every time we observe them anew, we get a new surprise: in one place, the animals show up every day; in another, they appear only in good weather—or only when the weather is bad. Where do they go when we can no longer observe them? Whenever the surface is even slightly troubled, their presence becomes difficult to detect. Sometimes, they seem to like enormous swells in the depths off a rocky coast; at other times, they daily navigate delicate passes, hunting between sandbanks where the bottom emerges at low tide. In the straits of Charente, the water is a maximum of 30 meters deep, and the sea is often calm. Nothing is fixed, nothing is definitively known. Henceforth, any report about dolphins, be it more or less scientific, will not only document the particular species of dolphin under observation, but also note the place and date in order to draw conclusions relative to this specific time and place, and avoid deceptive generalizations. Experience has taught us to suspect the certitudes of the past.

If dolphins are "ambassadors" for anything, they are certainly role models for modesty. This is an innate quality in many children, and it is strictly cultivated on the *Fleur.* Like adults, children want to have the answer to everything, but they still know how to ask questions and are open-minded enough to come up with new dreams when the old ones no longer fit reality.

When we began diving with the dolphins, I asked the children to forget all received ideas about the animals they were about to approach, and to act as if they were the first humans ever to pay them a visit. The interns on the *Fleur* are no different from any other lover of marine mammals—they want to understand everything, but first they are there to observe, ask questions, and bear witness, while accepting that they don't have the answer for everything. They assume nothing and hope for everything. This is perhaps the correct definition of the "scientific method."

In the Azores, our observation always begins on deck: we have to locate the cetaceans after having passed the night circling in the zone where they are found. At dawn, a member of the team charged with locating the animals climbs the mast to the crow's nest, while the divers designated to accompany me get ready: flippers, masks, and snorkels all are near at hand. As soon as the sun's first rays sparkle across the surface of the ocean, we usually sight the animals, and the captain heads out toward them.

One morning, we glimpsed furtive shadows flitting to the starboard side. I picked up the binoculars and directed them toward a group of slender torpedo-like shapes that seemed to be pursuing a school of flying fish. They were Atlantic spotted dolphins, about 2 meters long, which are easily confused with young bottlenose dolphins. Outside of the water, it is difficult to distinguish the little spots that speckle their bellies, and later, as they grow older, their steel-gray backs as well. They also have narrower beaks than those of the bottlenose dolphins. Lively and playful, they are also voracious hunters, very partial to catching squid in the depths or chasing surface fish while swimming in the company of albacore tuna. Furthermore, they are well armed: their jaws are adorned with as many as a hundred and sixty little conical teeth.

I asked the children to get ready while our skipper reduced our speed. Would the dolphins continue on their way, or would our presence incite their curiosity? For a few seconds, we saw nothing more, then suddenly, a dozen dolphins passed by us, this time to port side. I was unable to say whether it was the same group, but they showed no sign of being upset, and even less of aggressiveness. The *Fleur* seemed to fascinate them; they kept changing speeds, allowing us to catch up, then taking off again. It was classic dolphin behavior, but for some reason I had the feeling that we were in for a surprise.

"In we go!"

The names of two girls had come up when we drew lots: Morvana, from the Gulf of Morbihan, and Siobhan from Cornwall. Both of them were twelve years old. I ask them to be very attentive, because I had the feeling that the animals could be more numerous than we thought. We'd seen ten of them, but there were probably two or three times as many. Each time we dive, I accompany only two children. This is first of all for safety's sake, so I can keep an eye on them and be able to help in case a problem comes up. Diving without a tank in the middle of the Atlantic, even if it's undertaken with all the necessary precautions, all the same has nothing in common with taking a swim near the beach. But we also limit our numbers out of respect for our hosts, in order not to disturb them.

After climbing down the ladder over the hull, we gently and silently slipped into the water. "Going to see the dolphins" by jumping overboard with a big splash was out of the question.

"Of course," said Morvana, "I would never enter the house of someone I didn't know, kicking in the door without warning!" Her spontaneous remark proved to me that the children had understood how to go about their upcoming encounter.

The dolphins were still there. Usually the young ones, who are the least wary, behave like curious, alert, carefree adolescents. They can be crafty as well, because sometimes their conspicuous presence near a boat is designed to detract attention from the rest of the family. The females and their young thus remain apart. Such was not the case this time, for the pod came back to us when we'd barely done a few strokes. The two girls instinctively grabbed my hands, one on the left, the other the right, and we floated, immobile. The dolphins filed under us, at a depth of less than three meters, then a male turned on his side to observe us. The three of us still were floating, all holding hands. We were wearing colorful swimsuits, flippers, masks and snorkels. What a strange-looking six-legged creature we must have seemed to the dolphins looking at this spectacle!

In the beginning, the eye of the dolphin was like that of any other land mammal, probably comparable to our own. Because their eyes were adapted to see on land, when dolphins became aquatic, their vision would have been blurry. In the water, the image of an object no longer is formed on the retina, but behind it. Human beings correct this problem by wearing masks which keep our eyes in contact with the air, but cetaceans developed eye muscles over the course of their evolution which are strong enough to change the form of their lens. This became almost spherical, like that of fish, and the eyeball flattened out as their sight adjusted to a marine environment. By thus bettering their capacity to adapt, dolphins can see just as well under the water as above the surface.

Seawater is particularly corrosive, and a dolphin's eyes are protected by a thick, gelatinous mucus continually secreted through special glands. It covers the entire cornea, creating the impression that the animals are crying when they are out of the water. In fact, they would be hard-pressed to shed a single tear, for dolphins do not have lachrymal glands. They do have eyelids, which have no lashes, and which they close for the most part when they sleep.

Dolphins, like all other members of the tooth whale family, essentially have a one-eyed vision of the world. Ordinarily, as their eyes are situated on the sides of their head, they only see with one eye at a time. That said, the smaller delphinids' lower field of vision is binocular for about 20 degrees, which permits them to have stereoscopic, or three-dimensional, vision. Thus, when they wish to distinguish us clearly when we are out of the water, they position themselves vertically with their heads above the surface, and look at us by directing their gaze beneath their jaws.

Light waves travel poorly beneath the water, and darkness rapidly increases with the distance from the surface. We have learned that dolphins possess a reflective shield situated behind their retina, which facilitates night vision by reflecting light like a mirror. Species like beaked whales, sperm whales, and certain pilot whales, which dive to the depths of the ocean hunting squid, have weaker vision than the small delphinids, but seem able to detect the luminescent organs of their prey at a depth of 200 meters or more. Humpback whales have extremely mobile eyes, whose pupils dilate greatly in the darkness to capture the least ray of light, but contract into tiny slits whenever the whales surface. Other species living in troubled waters, like the Ganges River dolphin, have minuscule eyes with no lens. The diameter of their optic nerve is reduced, and the optic center in their brain has regressed to an extreme degree. The sense of sight does not prove all that useful in the muddy water they inhabit.

For a long time it was thought that dolphins saw correctly underwater, became myopic once they were out of it. In fact, their vision is much more effective than we imagined. Experiments with animals in captivity have proven that a bottlenose dolphin swimming beneath the surface can perfectly pinpoint an object 5 meters above his pool, and hit it every time he leaps. Dolphins can hence correct the variation in the refractive index caused by the change from water to air. Without this ability, they would systematically leap to one side of their target. This supposes an ability to adjust their sight which is greatly superior to our own, according to a process whose details we still are far from mastering. All the same, we know that their corneas have an irregular surface, which probably facilitates the adjustment from water to air, their eye muscles permitting variations in the curve of their corneas. We also used to think that dolphins could not perceive colors, due to the rarity of cones on their retina, but experiments with dolphins in captivity have begun to show the contrary. In other words, we still "see darkly" what it is they see.

The spotted dolphins that came to meet us were decidedly very curious. One of them began to swim in circles around us. As I was observing him, I felt one of the girls give my hand a hard squeeze. I looked down through the water and discovered my intuition had not betrayed me.

Dolphins were around us everywhere! For a split second, I thought we had plunged in the midst of a living flood where the currents, instead of carrying us off, swept around our bodies without touching us. That morning, the water was very clear. We could see a good 30 meters down—and as far as the eye could see, there were animals. Dolphins by strata, by layer, and we were right in the thick of them! No doubt there were others, too, swimming in the depths beyond our view. As the horde swam beneath us, the water they displaced moved all over our skin, giving us little caresses. We could hear them as well, whistling and making "clicks." I think that the entire aggregation numbered over three hundred, but I confess to not having counted. I was content just to admire them. How many times had I dreamed of finding myself in the midst of dolphins—and here I was! It was not a question of a nomadic movement, as spotted dolphins in the region of the Azores stay in the same place all year long. We were witnessing some sort of daily assembly. Pods and larger groups were joining up again after having gone about their separate ways, dispersing for a while to pursue their own activities.

Full of wonder and yet distraught, my two young companions searched out my gaze, looking for reassurance. I smiled at them, and would have willingly given them a comforting pat, if they had only let go of my hands for a moment! This was totally out of the question. They weren't really frightened, but all these creatures swarming around made their hearts beat so quickly they hardly knew what to do. Bit by bit, they got hold of themselves, and I heard them talking through their snorkels. No doubt they were expressing their joy and amazement. What they said was of course incomprehensible, but the dolphins nearest to them reacted to every sound: moving briskly away, they veered off at a sharp angle, or dived into the depths, trailing others behind them. It was as if the sounds the girls were making surprised them, causing their members to scatter. But soon they were back again.

And then, one of the girls cried out through her snorkel in a real fit of joy. In a split second, three hundred dolphins panicked and disappeared. We remained where we were, as if suspended in a vacuum. There they had been, so numerous they almost brushed against us, streaming by one after the other as if they were on a conveyor belt. Vanished! That was the last we saw of them for the entire day.

Later, on the deck of the *Fleur,* I remember secretly observing my two little divers. I can still see them now, sitting apart from the group, one up near the jib boom, the other at the foot of the companionway. Two misty-eyed dreamers. I know they were reliving those magical moments. Earlier on they had talked with their friends, gesturing animatedly as they recounted their extraordinary adventure. Still later that night, they would write and draw in their logs for hours, trying to capture their sensations and emotions. But at the moment I came upon them they were alone, lost in themselves and their emotions. I'm sure their hearts began to beat a bit more quickly as they remembered the spotted Atlantic dolphins stirring up the ocean all around them. Such rare emotions abide with us for a lifetime.

# IV

## *Swept Away*

*Floating.* © Margaret McCarthy

# Life at the Cauldron

*John P. O'Grady*

Just after sunrise on a day in May not long ago, in a hot spring pool near the South Fork of Idaho's Payette River, a man's body was found floating face down in the water. He had been shot once in the head. The incident occurred in the middle of nowhere, so police investigators had a long drive to get there. Little evidence was recovered at the crime scene: no eyewitnesses, no signs of a scuffle, no bloody footprints leading away from the body. Police did talk with some kayakers who were staying at a nearby Forest Service campground. They had been down at the hot spring on the previous evening as late as eight P.M. No body there then. Now there was, and the kayakers were upset. This was the wilderness, they said. Things like this aren't supposed to happen here.

The police bound off the area around the hot spring with a yellow plastic ribbon. "We're setting up a perimeter," an officer explained. Onlookers nodded their approval. Printed on the yellow ribbon in black letters was a message: "Police Crime Scene—Do Not Cross." The message was repeated continuously along the length of the ribbon, for more than a hundred yards. Since this was a perimeter, the ribbon came back upon itself and formed a rough circle on the landscape. The message became a kind of infinite loop. The onlookers took comfort in seeing police authority stretched in this way around such a tainted scene. The order of things had suffered a deep gash, but now a tourniquet was firmly in place.

The investigation proceeded. Indications were that the man had been killed right in the hot spring pool. He had identification on him but no money. How he wound up in the pool, who killed him, and why were formidable mysteries. It was as if, standing at the bottom of some colossal billboard, you looked up and tried to make sense of it. The upper parts would appear smaller and the lower larger than they should, while all of the colors and lines would swirl in clouds of confusion. The totality of the image does not reveal itself to one standing too close. Only backing off will bring the big picture into focus. Could be that it's a man hanging miserably on a cross, or perhaps an oversized Guernsey wanting to know if you've got milk, or maybe it's an extra-tall circus clown peddling hamburgers. In any case, when it comes to a mystery, achieving the proper vantage point is critical.

The first break in the case came a few days later when an abandoned car was found in the city. It was a blue Ford Escort. Somebody had parked it behind two huge dumpsters in a hotel lot, not easy to see. "Whoever was driving that car knew it was hot and stashed it," said the cop who found it. In the weeds along the edge of the parking lot, investigators discovered a crumpled bumper sticker. It said, "I Believe in Dragons, Good Men and Mystical Things." It had been taped in the rear window of the Escort, and somebody had removed it. A detective said, "That bumper sticker was taken from the car, thrown in the dumpster, and the wind blew it out."

As it happens, both Escort and bumper sticker belonged to the dead man. When the police searched his apartment for evidence, they found among his effects a number of books on marketing strategy, customer behavior, and witchcraft—as well as a strange homemade mirror, the surface of which was painted jet black. It reflected nothing. "What the hell is this?" asked the cop who found it. Nobody could say. They took it away to the police evidence warehouse, along with anything else that might have some bearing on the case. Fingerprints other than the victim's were found all over the apartment, and they matched sets taken from the car. Soon two suspects were in custody. Over the next few days, more details were flushed from cover, and a story unfolded in the newspapers. It went something like this.

The dead man's name was Kelvin. He was twenty-nine years old and had been studying marketing communications at the university in Boise. He lived in an apartment close to campus. Although raised in the Mormon church, in recent years he had become enchanted by other things. In the months prior to his death, he had been studying witchcraft, but not at the university. He was receiving private instruction. Names were never revealed.

People who lived in Kelvin's apartment building described him as somebody they didn't really know. He was a discrete character; nothing in his world seemed to overlap with anything in theirs. One of his fellow citizens—you couldn't really call them "neighbors," as the only interest they held in common was that the city removed their trash once a week—told a television news reporter, "There was something strange about him. I can't describe it. People come and go, and that's pretty much it."

On that night in May, Kelvin drove his car from Boise into the mountains. He was accompanied by two young men in their late teens. News accounts referred to them as "transients," but they were friends of the victim and well known around town. The two young men each had a long record of petty theft and writing bad checks. Kelvin somehow befriended them, even though he himself had never stolen anything or written a bad check in his life. His mother described him as "a good-hearted man."

He often drove the two young men around in his car because they didn't have one. Together they ate a lot of fast food and smelled like it. Sometimes he would let them stay with him in his apartment. The three were part of a larger group that was caught up in Dungeons and Dragons, a role-playing game that skirts the line between fantasy and reality. At some point, Kelvin told the two young men about his witchcraft studies and showed them his magic mirror. They were fascinated.

One day in the midst of a Dungeons and Dragons game, the two young men noticed a receipt lying in Kelvin's apartment. It was for an $8,000 student loan. The two young men figured this was a lot of money, so they decided to steal it. The only question was, How? The money was in Kelvin's bank account, and they needed a plan to get at it. The magic mirror was lying nearby, so they consulted it—perhaps in the hope of finding a co-conspirator among the disembodied—but they couldn't get it to work. Then the two young men noticed that Kelvin kept a couple of hunting rifles in his closet, so these became part of the plan. Things grew apace, like cheat-grass across the western range.

Here's what they came up with. They would lure Kelvin to a remote area, force him at gunpoint to give over the access code to his bank account, and then kill him. After that, they would return to the city and withdraw the cash. "This was their way," a detective put it. "They were going to split it $4,000 each." The two young men were deep into a new game.

Around 10 P.M. on that fateful night, they suggested to Kelvin that they all drive up to the hot spring. It was an odd hour for such things, but for some reason Kelvin was drawn to the idea. You might say that at this point, his life had become a deep basin of strange attractions. There are those who say that he just didn't pay enough attention to evil, for if he had, he never would have spent time with these guys. Experience would suggest there are certain souls who are like cheap motels where all kinds of atrocities find ready lodging, and here were two young men with "No Vacancy" signs burning in their eyes.

They all got into the car and headed north. Their destination lay more than an hour's drive through the dark woods of Idaho. High mountains added to the darkness. By the time they arrived at the hot spring parking lot, the waning moon was struggling to rise over a barren ridge. It looked like a shipwreck. There was just enough light to make out the long descent of stone steps to the hot spring pool. Kelvin led the way. Although he had never been to this place before, it seemed strangely familiar. The ponderosa pine made the night air smell like an ice cream cone. A coyote barked in the distance. The river churned through the gorge below.

Kelvin didn't notice anything was wrong until he was sitting comfortably in the warmth of the hot spring. Only then did he see two sinister images reflected in the black waters of the pool—a couple of bad ideas suddenly manifesting in the mind of God—and each one pointing a rifle at the image of Kelvin's head. Had he attended carefully to the semblances in the mirrored waters, he could have witnessed the terror blossoming in his own eyes like a pair of phantom orchids.

The jaw of one of the images started moving up and down as if to speak. It was like a dream. Kelvin heard a voice demanding the access code to his bank account. The voice seemed far away, as if coming from the moon. Kelvin pleaded with the images for his life. He gave them what they asked. It wasn't enough. A sudden flash and the whole reflection shattered, as if something heavy had just been dropped into it.

When the two young men got back to the city they discovered that Kelvin had given them the wrong access code. Now there wouldn't be any money. They were really angry, and they cursed their victim savagely, revealing the truth in the old saying that offenders never forgive. In their rage, they ditched the car but forgot to wipe it for fingerprints. Soon after that, they were in jail. Although both confessed to plotting the crime, it's not clear which one actually killed Kelvin. Only one shot was fired, but each young man accused the other of pulling the trigger. The rifles had been thrown into the river and were never recovered.

The newspapers kept hinting there was something strange about this case. It started to come out during the preliminary hearing when the prosecution asked the sixteen-year-old wife of one of the two young men if her husband was into witchcraft. She replied, "From my understanding, he was into . . ." She never completed her thought because her husband's attorney objected to the question. He knew a Pandora's box when he saw one. The judge sustained the objection. There was already plenty of evidence to bind the two young men over for trial.

As more of the details of the Hot Spring Murder (that's what the newspapers were calling it) came to light, especially those pertaining to witchcraft and magic mirrors, I found myself reflecting on my undergraduate days at the University of Maine. It was a time when weirdness hung in the air thicker than the pot smoke in the dormitory hallways. I was taking a class on horror literature from the writer Stephen King, and for a term project I decided to spend a night in a haunted house. All I had to do was find one.

I heard about a place down in Bar Harbor, so I went to investigate. Although there were lots of stories about this house—most of them involving a headless sea captain come back to claim a hatbox full of love letters—finding the place proved more difficult than catching its ghost. People would say things like, "I hear talk of that house, but I've never actually been there myself," or, "Oh, yeah, I was there once in high school, but it was at night and I was with a bunch of friends and we were kind of drunk. I think it was over by Otter Creek, but that might be some other place." And a surprising number of people in town seemed to have the same "friend of a friend," a courageous fellow who tried to spend a night in this house but was chased out by a ghostly set of limbs and a torso that suddenly appeared and then tumbled down the steps, as if spilled from a children's toy chest or something, only to assemble at the bottom as the headless sea captain, which then gave chase to the courageous fellow, who fled the scene and indeed the town of Bar Harbor, for he had not been spotted there since, though his hair is said to have turned white for the experience and that's how you will know him if you ever run into him.

Over the course of my inquiries, one name kept coming up as someone who might be able to help. "Go see Melissa," they said. "If anybody knows where that house is, she does." As I soon discovered, this Melissa was a woman my age who was "into" witchcraft, her specialty being the manufacture of magic mirrors. This may seem an odd occupation nowadays, maybe even a little quaint, but it was the sort of thing you did in Maine back in the seventies if you didn't own a big summer "cottage" and you had gotten tired of working as a laborer in the gardens and country clubs of the rich or as a deckhand on their yachts. It was either that, or you married one of them, or you robbed their places in the winter. There just weren't many options.

Melissa had a little shop in downtown Bar Harbor, so I stopped in for a chat. I got to know her somewhat over the next few weeks. She had hair the color of a fire-bush in the fall, and she wore these long flowing dresses all the time and shoes hardly ever. It was a look that in college we called "organic." You could almost see the bits of granola shaking out from the folds of her peasant skirt as she walked down the street, or imagine the little flock of birds following in her train feeding on what bits of fruit and nuts might be gleaned. "Watch out, O'Grady," my friend Westphal warned me when I told him I had been speaking with her. He had gone to high school with her and knew something of her history.

Tradition had it that on her mother's side, Melissa was descended from a woman who was executed in the Salem witch trials. This curious bit of family lore—no ordinary skeleton in the closet—had been a great source of embarrassment to Melissa while she was growing up, but in college she took a course in women's studies from Professor Phoebe, who convinced her that witches are an oppressed minority. Having a genuine Salem witch hunt victim for an ancestor could be a fabulous boon, the professor convinced her, so Melissa embraced her heritage and turned it into a business.

She dropped out of college and opened a shop that she called "Mirror, Mirror on the Wall." She sold nothing but genuine magic mirrors, each of which she made herself. Assembling them was an elaborate process, requiring carpentry skills and lots of time and labor, plus it all had to be done at night when the moon was waxing, because this is what endowed the mirrors with a special power. Finally, according to what I heard, each mirror had to be "jump-started" by being dipped in a certain tide pool in Acadia National Park at midnight when the moon was in Pisces, an aspect of production that, had the Park Service known about it, would have led if not to Melissa's persecution then certainly to her prosecution. Park Service administrators have an aversion to occult practices' taking place anywhere on government property. Such goings on, they declare, stigmatize the resource.

So there I was in the witch's shop, with its abundance of strange mirrors hanging on the wall, each one looking like a blemish on the face of all things bright and reasonable. Trying to hide my discomfort, I introduced myself and casually asked about the haunted house and its headless sea captain. The witch laughed at me.

"Don't you believe in ghosts?" I asked.

She laughed again and said she had seen far too many to believe in them. "That's just a story," she added, with a hint of that pity usually reserved for the easily duped. Meanwhile my term project was vanishing faster than its ghost.

"Well, then, tell me about these mirrors," I said. "How are they supposed to work if you can't see anything in them?"

The witch explained that these were *magic* mirrors, not for ordinary eyes but only for those who wish to delve beneath the surface of things. Each magic mirror, she assured me, was a kind of mental door into a marvelous world.

She lifted a hand-held mirror from behind the counter. "Here," she said, "try for yourself. Think of it as a gate for your mind to go through. When it opens, you'll have a way to get out of your normal life." Melissa, I quickly came to realize, regarded ordinary consciousness as something that needed to be escaped. For her, the word *normal* had all the connotations of Alcatraz.

For my part, I had concerns about doors like this. If they're locked, I figure it's usually for good reason. But at that point, time was running out for my research project, and I needed something to replace the haunted house. I thought maybe I could write something about these magic mirrors. Later, when I ran the idea by Steve King, he was all for it. He already knew enough about haunted houses and was game for a look into other things.

Nervously, I accepted the mirror from Melissa and had a look-see. Nothing.

"What am I doing wrong?"

She was more than willing to set me aright. There is a part of the mind—she called it a third eye—that can see the images that dwell deep in the magic mirror, but first that eye must be opened. Let me recount what she said, as best as memory will permit of long ago and more chancy days.

Everybody knows that a mirror is supposed to reflect things, but less apparent is its capacity to *absorb* the images it captures. The dull black surface of a magic mirror reflects nothing, and the point of this opacity is to make it easier to draw forth the images that have already been lodged away in there. Think of it: the mirror in your bathroom or in your car, the storefront glass you pass on the way to work, even your beloved's eyes—all these are not just hurling back the accidental rays of your sweet self, but they are also soaking them in. An image cast into a mirror is like a deposit made in some spectacular treasury, a veritable Fort Knox of the soul—something like memory itself. Not only that, but each mirror is linked to every other mirror—past, present, and future—so that by having an "account" in one, you actually have access to all the others. It would be something along the lines of a vast banking network that cuts across space and time. The magic mirror is your passbook.

A few days later when I returned to the shop for a "mirror session," Melissa sat me down at a small table in a dimly lit back room. She set a mirror—"I chose it specially for you"—in a small tripod on the table. "Concentrate," she said. "Gaze into the mirror. Don't be filled with distractions. Take time. Relax. Empty your mind. Look at it. Look through it. Let it open up. Have no expectations, no demands, no agenda. Take time. Relax. Empty your mind." Then she left me alone with all that emptiness, closing the door behind her.

A magic mirror, in the broadest sense, is really any object whose surface lends itself as a projection screen for the unconscious. That means just about anything will do. Most assuredly, the world itself is a magic mirror. That includes your family, your house, your car, your religion, your job, all the grudges you harbor, the ambitions you nurture, the games you like to play and the games you like to watch, as well as that empty beer can somebody tossed on your lawn last night. It also includes the old woman in the supermarket aisle who, finding all the cereal boxes too big, shouts to no one: "Don't they know millions of us live alone? How can I eat all these corn flakes by myself?!"

As might be expected, the permeable style of consciousness cultivated by habitual use of magic mirrors has its drawbacks. Melissa admitted that she had become so adept at seeing images in her magic mirror that she now saw them in all kinds of "inappropriate places," like swimming pools and glasses of water in restaurants. There were some days when she could hardly bring herself to walk down the street for all the house windows glaring at her. When I asked her why she didn't just flip off those unsolicited images, she said, "It's never wise to be rude to them. They react maliciously."

My mirror session caused me a few problems of my own, even though I was never able to see anything in it. That night I had a disturbing dream that recurred for next three after that.

In the dream, I'm standing outside this huge red-brick slaughterhouse in Sioux Falls, South Dakota, or someplace else, and there's this high-pitched whine coming from inside the building, which sounds like a hundred giant band saws all going at once. I don't see them, but I know that inside the slaughterhouse are thousands upon thousands of trembling cows being herded, one by one, to their bloody band-saw deaths. Outside, where I am, it's sunset, and suddenly a dark swarm of bees, thick as an inversion layer of mortuary smoke, comes rushing in from the bleeding horizon to blot out what remains of the sun. But when the swarm gets closer, I see it's not bees but a huge flock of passenger pigeons buzzing like bees, and they start gathering over the red-brick slaughterhouse, thicker and thicker, until, one by one, each bird bursts into a ball of fire. Meanwhile, inside the slaughterhouse, the cows continue to be herded, one by one, and for each bloody band-saw death that occurs, a passenger pigeon explodes in flammable witness. That was always the point when I woke up.

I told Melissa about this dream and asked her if it had anything to do with the magic mirror. She said not to worry; it was just warm-up for bigger things. Steve King, on the other hand, suggested caution, casually mentioning that spirits are known to flock to sacrifice because they feed on spilled blood. "Maybe something *bad* came through that mirror, and now you're dreaming about it. I'd watch it if I were you."

Oh, I didn't like any of this. Right then, I swore I would never look into another magic mirror. And just to play it safe, I became a vegetarian. That was over twenty years ago, and I've stuck to it.

Today when people ask me why I don't eat meat, they presume it's some sort of moral thing. Usually I tell them I just don't like the stuff, that it's a matter of taste, not morality. I don't tell them my real reason for avoiding flesh: *fear.* I also don't mention how I bypass the butcher section in supermarkets like some people steer clear of bad neighborhoods, or how, when I'm driving down the neon thoroughfare of America, I make sure not to stop or even look too long at all those McDonalds and Wendys and Burger Kings—for fear of what I might see swarming through the grease clouds that billow from the kitchen vents on the rooftops and spin off in ghostly threads to attach like phantom lampreys to cars that pass through the Horned Gate of the drive-through. I never mention this kind of thing. Philosophical talk about the nature of sacrifice is best avoided in polite company.

One night, a few weeks after Kelvin's body was found, a Forest Service ranger noticed some strange goings on down at the hot spring and decided to investigate. When he descended the stone steps to the pool, he found candles burning everywhere and a dozen or so men and women standing holding hands in a ring around the hot spring. They were all naked. In and of itself, nudity at a hot spring is nothing uncommon, but these people were chanting some kind of strange prayer while one of them dangled a mirror from a string over the pool. The forest ranger thought this suspicious. When he confronted them, they said that were doing a "working."

"What's a working?" the forest ranger wanted to know.

The naked people stood there quietly for a moment. A couple of them shuffled their bare feet. The candlelight flickered on their bodies. Then one of them said the soul of the man who was killed here was trapped in the pool. They were using the mirror to catch and release it.

The forest ranger told them they had to break it up.

Since timber production has fallen off considerably in Idaho in recent years and most of the mills are now closed, the U.S. Forest Service is trying to redefine itself. Recreation is replacing wood as the main product coming from the public lands, and the agency is putting a lot of money into its "interpretive programs." That means lots of new signs—or "signage," as they like to say—are being put up at the more popular sites such as hot springs.

So if you visit a particular hot spring near the South Fork of the Payette River, you'll find some very big signs in the parking lot. They tell you about "Life at the Cauldron." On good authority you will learn that the flora and fauna associated with the hot spring environment are "supported here but not elsewhere." If, during your "hot spring adventure," you look around carefully, you might spot the narrow-winged damselfly (sometimes called "fire-breeding dragon"), or you could see the hot springs snail, or the soldier-fly, or maybe some panic grass. If you're really lucky, you might encounter a "rare orchid that in Idaho grows only near hot springs." The signs, however, say very little about the most common species found around here, whose highly visible traces include empty beer cans, stumps of candles, and soiled underwear tangled in the artemesia shrubs. About the only thing you will learn is that "some early settlers in this region believed that such hot, smelly water associated with the spring escaped from the 'hellish' interior of the earth."

If you come to this place hoping to learn more about the Hot Spring Murder, you will be disappointed. The signs in the parking lot say nothing about it. Nobody you ask in the campground will know anything. You may see a couple of overturned trash cans and some garbage strewn about—the work of a local bear—but you won't find any remnants of yellow plastic ribbon. The long descent of stone steps may seem familiar, but you won't know why. When you get to the bottom, you will find nothing telltale about the hot spring itself. And if, at last, you take a long, deep look into the river, now settled into its placid flow of late summer, you may see your own reflection scudding along amid the eddies, but these waters will not give up their secrets.

You could stop in at the U.S. Forest Service information station a couple of miles down the road and make some inquiries, but don't expect to be received warmly. The Forest Service wants to manage the interpretation of the forest in the same strict and scientific way they used to manage timber. Unregulated stories such as the Hot Springs Murder are like bark beetles or forest fires, a counterproductive side of nature that is best eradicated or brought under control. When a forest ranger talks about rumors spreading like wildfire, you can bet he is not just speaking metaphorically.

I stopped in there myself last week to obtain a little information. The ranger was happy to provide me with all kinds of materials pertaining to the hydrology of hot springs and where I could park my RV if I had one. He spoke glowingly of all the "recreational opportunities" in the national forest and generously offered to tell me where the best fishing holes were. He was wearing a ranger hat and doing a good job.

Then I said, "So, what's this I hear about naked witches down at the hot spring?" It was an inappropriate question. He shook his head grimly and looked at me as if I had just committed a crime, and then said, "It's a real shame that resource has become stigmatized."

## There's a Lot of Room

*Eva Salzman*

Art was a lot taller than Petra, and dark and handsome, so it seemed a privilege to be kissed by him, yes a privilege.

They'd walked down Bay Avenue, across Montiak Highway toward Vale Road and then doubled back to where they had started, the Town Dock. It was the circuit. There wasn't anything else to do but go around like this. Petra was too nervous to talk—her heart was hammering away and there was a tingling in her stomach—and Art didn't talk either.

Her deep crush had started at the Hamden Bays Square Dance. She couldn't believe her luck when he asked her out—by which was meant she was to sneak out of the house at night and meet him down at the town dock. There would be someone else for her sister June.

Back at the Town Dock, he kissed her.

His arms were wrapped around her shoulders and he wasn't moving much. In fact, he wasn't moving at all. He planted his large, somewhat dry mouth over hers, and there it stayed.

It's not that she minded that, of course, but, after a while, she found herself waiting for something else to happen, thinking he might tilt his head the other way for variation, or maybe suck or push a little with his lips, or even introduce a tongue. But the activity, as Art apparently understood it, was completely stationary. Nor did his hands move down her back or try for her large breasts—which June had insisted would be Art's primary goal.

June was at this very moment stuck with Art's friend, a puny little kid—a kid, the way that Art was not. Art was really into man territory with his height and girth, if not his kissing technique.

Petra had no idea what had happened to them right now. They had melted away into the dark night and now there was just this kiss.

Eventually, her heart began to slow down, until it reached its perfectly normal, unexcited, pedestrian pace.

At first, she'd kept her eyes shut, as you were meant to do. But after some time had passed she had a peek. Through her lashes Art's face loomed up, frighteningly close, so large and round it reminded her of a balloon, or a moon. She was shocked by her own realization that it really wasn't all that attractive from this perspective.

So she closed her eyes again. Her face was probably also like a moon, but Art wasn't looking, or not while she was looking anyway.

All her concentration was on this large, dry mouth like an "o" fitted over her mouth. No biting (some boys did that, or they caught you with their dental braces), no sucking or gentle pursing to punctuate the activity. She began to wonder if it would ever end, what exactly would signal the end for him? What would make him stop?

She couldn't possibly end it herself, god no. It was his kiss after all—he had bestowed it and she was the taker. She'd feel stupid, ungrateful, embarrassed. So she just stood there.

It was so dark that at one point she slightly staggered, losing her balance. He just moved one leg slightly to improve the position, and his open cave mouth enveloped hers again, so everything could return to normal. She liked his smell. He smelled like wood. For a second she had a fantasy that he was wood, that they were both carved wood figures, locked in kissing position, posing for . . . posing for . . . what were they posing for? Maybe someone had artfully placed them there. Who? Why? Artfully. That was funny.

The smells of the beach came and went, little sharp draughts of salt and seaweed and crab. She could hear the light lapping of waves on the shore, the gentle creak of a moored boat, the occasional bumping sound of the hull against the dock. A light breeze brought the sharp tang of the nearby marsh—an intriguing rotten smell of mud and dank pond. Dank pond. That was vaguely unpleasant, not a word she should be dwelling on at this particular moment.

Eventually she drifted further and further away until she was home again, the very place she had escaped from that night, for these irresistible illicit purposes. Her mother and father were asleep in their bed. She entered their room and for the first time imagined them sleeping together.

Maybe her parents had kissed like this, or maybe they hadn't, maybe they'd kissed another way. Maybe her father was a lousy kisser. Was it possible for two people to stay together if one was a lousy kisser? Did he kiss other girls the same way he kissed her mother?

Now she felt like she was inside a coffin, the world had shrunk so much in the pitch black. She'd lost her perspective so utterly she might as well have been lying horizontally underground. She'd forgotten what she was doing, this activity so far from the subject of corpses and decay. Of course it didn't matter if she did forget—this kiss would just go on anyway, with or without her.

She'd lost all sense of direction and decided to occupy her time playing a little game with herself, to figure out her orientation. For this reason, she now positively wanted the kiss to end, to prove her calculations correct. Also, she was missing the considerably satisfying sight of June having a gloomy time with a pint-sized boyfriend.

But still it wasn't finished. Maybe this was the right way of it, kissing sort of like meditation, with your mind not necessarily on the matter at hand. After all, should one have to concentrate on the actual kiss? Then it might be more like work. Or maybe it was like an antechamber, or a large cathedral-like place itself. There was so much room in it, she'd never realized.

But why should the duration of this kiss also make it disappointing, make her feel robbed and deflated?

She knew now he had no intention of going any further. Besides which, if he did decide to progress in the usual manner, she might end up with this large, dry hand planted on her breast for an equally absurdly drawn-out epoch, unmoving and spread out over her flesh, more like a brassiere than the curious, groping, desirous fingers of a hungry adolescent. And that might not prove to be much fun either.

So it was a considerable surprise when finally he closed his mouth and stepped back.

Then they walked back, arms wrapped around each other, and said goodnight. Once Petra and June had been dropped off, had softened the screen door's slam and made it safely up the stairs to bed, Petra found herself getting excited again at the thought of Art.

Of course she would have to tell her friends all about the kiss, about the way it was, as she remembered it, since it might never come again, or if it did, she might not have the patience for enough time to pass. How would she ever recapture the thrill which had waned so dramatically?

She wouldn't. It was the meanness of memory that left her forever locked into it, a ghostly statue of the two of them entwined on the beach, for as long as she could remember, something so wonderful only in remembering. And it's not that anyone would ever know they were there, would only step through them on their way to somewhere else.

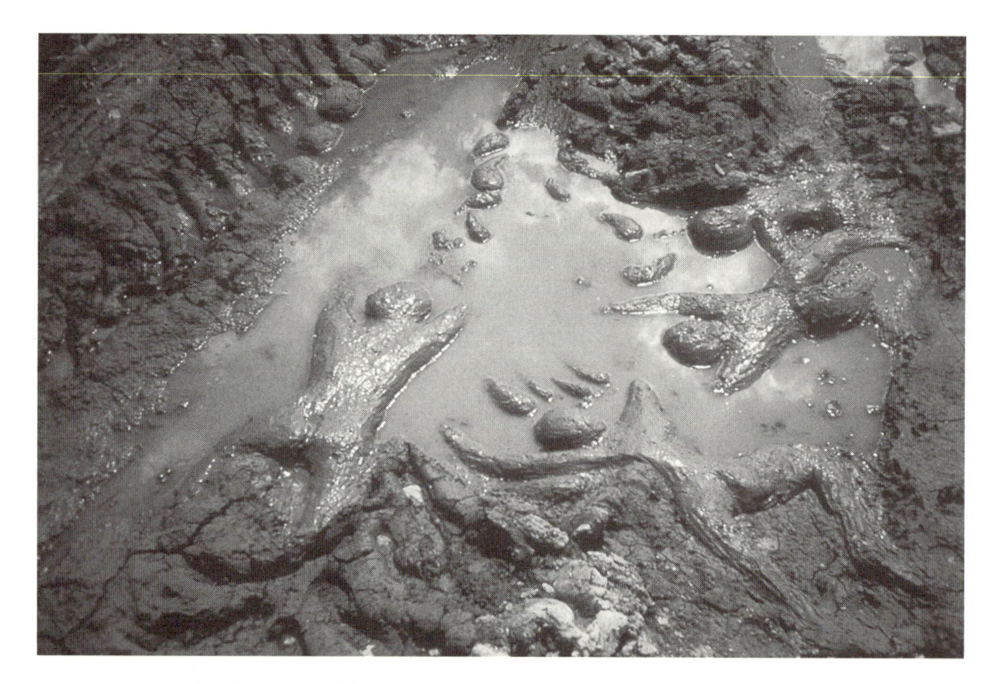

Monique Crépault, *Abercorn 3 (Afternoon #3)*, 1998, mud and water.

# Liquidation

*Tom LeClair*

## *I*

Everything flows, the ancient Greek said, from the river bank. Barging down the interstate, we tell you everything fails. Retail and wholesale, manufacturing and service, ingenious start-ups and old-line standards, the narrow-niched and the broad-based, the local and the international, businesses, companies, firms, conglomerates—they all fail. Margin shrinks, profits plummet, losses mount, and we dissolve the assets, turn movable goods into liquid money, transform trailers of objects into lines of digits on liquid crystal displays.

To compete with other road shows—monster trucks, heavy metal acts, wrestlemanias—and undersell local discounters, we're a tour de force, a sudden invasion and four-day display of surprise. We're force on tour, thirty high-cab Kenworths filling the right lane like a military convoy, tractors and trailers all the same gun-metal gray. From the two-lane highways and access roads, our closed nose-to-tail formation looks like boxcars—MIDWEST LIQUIDATORS, MIDWEST LIQUIDATORS, MIDWEST LIQUIDATORS—trundling toward some final depot. From closer up, the rest stops and weigh stations, the diesels' roar and smoke demonstrate that our over-the-road army-navy store carries every American service's surplus. At a steady fifty-five miles an hour, we're a forced march with all the products of forced sales. And to oncoming traffic, our day-time headlights show that, like a funeral procession, we're hauling the heaviest weight, the dead weight of failure.

Since 1973, when wages stagnated, we've seen affluence run its course. At home and abroad, the dollar declined, and America contracted. Even words failed. In the age of global competition, "foreign aid" became taboo and then an added "s" finished "aid" as goodness, stained giving with incurable disease. Before AIDS and before charity concerts, those extravaganzas that hyphened "aid" to defunct groups, we could quietly announce our arrival in your city. Before the Community Chest emptied out and the United Fund was plundered, we could subtly advertise our altruism with a minor misspelling: Midwest Liquaidators. Now we're forced to cast a major spell, come in big and come on strong, if we're to aid all we serve: the sinking entrepreneurs, the family concerns going under, the franchises drowning in debt, the corporations that can't be bailed out, and you, all of you who walk our aisles, survey the products in piles like the wrack of flood, and buy the goods we offer at savings only liquidation allows.

From hundreds of miles away, we saturate you with unexpected air power in Tuesday drive time. "Whump whump whump" the ads begin, helicopter blades reminding veterans and moviegoers of bullets from the sky. Then our traffic reporter screams over the noise: "Everything fails. The Liquidators are coming, the Liquidators are coming. Out past the loop, traffic is backing up." (The sound of downshifting, the surge of torque in a lower gear.) "Cars are lining up behind the Liquidator trucks. It's a mile-long caravan following the Liquidators to the arena." (The beep beep of happy horns.) "Bring your trailers and vans and pickups and empty trunks," the voice shouts and hesitates before identifying the appeal of returning armies, "collect the spoils." Then, over the returning helicopter whump, "Thursday through Sunday while they last, America's best deals on wheels." Finally, almost covered by the noise, a fading trailer: "Only once this yeeaarr."

Once a year every year, our tour of duty takes us to the Midwest's forty-five largest cities, a 1,500-mile jagged loop from our base in Middletown, Ohio. We appear in your city without warning. We're our own advance guard, "Marines of sale" our radio voice suggests, victors in the American price war, the kind of road warriors who'd put competitors in cement footwear. To give our show of force respectability, to show you we're not like gypsy roofers or those rear-door distributors who sell direct from stolen trailers, we rent the biggest buildings available—metropolitan arenas, coliseums with surround-around seats, small-city domes, homes of minor league franchises. We steer away from exhibition halls and convention centers, spaces designed to show off future success, low-track lighting burnishing the gloss of next year's boats and campers, mezzanine booths with high-definition screens to project trade show dazzle. With our marked-down goods we need to be at the bottoms of buildings that have no basements. We want the high space over our heads, the empty seats, canvas tarps covering courts, rough wooden flooring over ice. If stadium managers would let us, we'd spread dirt like the rodeos to remind our customers that Liquidators are the lowest link in the chain of sale, the chain of chains: the high-rise department stores dropping goods down to their outlets, the outlets dumping to the alphabet of marts, the marts dispersing the unsold to odd-lots and seconds shops, those one-room collections of ill-made and damaged objects in abandoned strip malls. If the goods change hands after us, it's underground, the underworld of flea market, trading stall, or garage sale.

On Wednesday we dolly in the crates and boxes, remove the merchandise, pile the containers in walls, and make a maze. We set our wares on the floor, fanning out irregular shapes—wooden duck decoys, coffee-makers, pillows—and stacking up rectangles and squares—socket sets, VCRs, and tool boxes. No shelves or tables or bins raise and organize our low-tide remnants. Narrow aisles coil and loop through the almost solid mass of solids. Hypermarket grids don't section, and suspended signs don't name our display of dense disorder—bathroom tissue stacked next to touch-up guns, Ninja Turtle backpacks spilling into Hocking microwave cookware, layers of industrial tarps across from stands of beer logo pool cues. Our design is unpredictable combination, the familiar scrambled into strangeness, a rapid succession of surprises whatever curling path you choose. Around the curve ahead, over the wall of brand-name boxes, or far across this huge floor are, somewhere, air ratchets next to wicker baskets, boomerangs sliding into surge protectors. Without clear sight lines or consumer categories, somewhere is anywhere, anywhere is everywhere, and the scale of our show seems prodigious, a Kenworth cornucopia, a feat of skill as well as strength. The banked seats of the arena, visible in the distance, are like the shoreline of an inland sea we've drained to reveal detritus that centuries have amassed in strange heaps and meandering folds, all present and waiting under this imaginary body of water the Liquidators have turned to air, this dead sea of failure.

Wednesday mornings we give away the catalog to our exhibition, a supplement in the newspaper, a booklet delivered to the doorsteps and apartment entryways of poor neighborhoods. For those who don't read, the flyer jams into its eight pages hundreds of two-inch black and white photos partly covered with yellow blazes and red prices. Crowded in among the pictures and fire-sale splashes are, for readers, facsimiles of brand names and brief descriptions. The cheap paper and lurid colors, the jumbled photos and blazoned warnings—"prices subject to change," "limited quantities available"—make the flyer an album of impermanence, throwaway catalog of a temporary installation never heralded by banners on downtown lightpoles. Although pages appear hurriedly composed, artlessly assembled like a hyperactive child's collage, they map the floor's maze. On the front and back pages and next to the margins is light reading, photos of men in T-shirts and women in brassieres, pictures of plastic bracelets and plaster knickknacks, come-ons for kids like baseball cards and barrettes. These goods line the edges of the floor, penny items for pockets, dollar buys for handbags. Moving inward from front or back, the browser finds double-digit prices—skateboards for $14.95, vanity mirrors at $11.95, bottle jacks at $23.95, boom boxes for the whole family, "only $79.95, compare with name brands." At the centerfold is the heavy reading, the heavy-duty items at the center of the floor: a Campbell Hausfield portable winch, $179.95; a forty-pound sandblaster system, $199.95; a six-inch long-bed joiner, $299.95; a Voltmaster 6,000-watt generator, $399.95. Garlanded by hand tools and housewares, games and adornments, these durables are the load we've toted, the lodestone to which our prized customers gravitate, leading their women and children inward—and downward if our center-weighted space seems like a bowl—to merchandise with force like the force that moved it, power tools requiring power to take them home. At the newsprint crease and coliseum axis, anachronism is the Liquidators' basic appeal: an all-male band of teamsters and stevedores bringing machines to fellow anachronists, men who still make things at home.

Wednesday nights we guide the customers from their vehicles with videotape. The radio broadcasts and newspaper spreads are the same in every city. The six and eleven o'clock TV ads take the local, eye-level point-of-view, documentary film that might have been shot by any customer with a camcorder: an outside view of the arena and parking lot, a moving, bumpy shot of building entrances, a slow pan of the jammed floor, the camera then winding through the aisles, glancing right, lingering left as one of those intimate home shopping voices recites the brand names, hushed accompaniment to the marvel of so much, the liquid sounds and fluid sentences preparing for the quick change we Liquidators undergo when, on Thursday morning, you follow the film and enter the show. Burly men who yesterday wrestled crates are sitting on folding chairs squeezed in among the offerings. Today, wearing loose-fitting smocks, these soldiers of bad fortune look as effete as museum guards. Raucous Wednesday roadies are as quiet as undertakers. You're surprised to find our power has been exhausted. Now we're as harmless as prisoners of war waiting for a handout.

You too have been transformed. Former spectators in this venue, you're out of your seats and down on the floor, all of you now suddenly athletes—men, women, and children walking where you've never been before, unfettered by ticket stubs and officious ushers, circulating freely where you've watched all the hometown heroes, moving where you want, ignoring if you wish the scattered spectators sitting still as their wares, passive observers of your motion, respondents to your desire and will. You take any path through the floor's field of force, wander the twisting aisles waiting for impulse, or search the piles for things you need. Sliding along like skaters in slow motion, towering over the floor-bound goods like high-rising hoopsters, you're the winners now. In the seller-buyer conflict we can never completely hide, now you're the ones with force. We give you the power of purchase, physical purchase, literal leverage, a place to stand and bend and lift, every shopper a shoplifter. Man, woman, or child, you reach down, pick up, and hold. You lean in, stretch out, and heft. You raise your arms, grasp, and weigh. Everything is within your reach, no sales cases or sales persons, no display models and backroom replacements. Everything is here, right here, up for grabbing, down for grubbing, everything almost moving with the motion that got it here.

Even auto batteries, stacked near the machines they start or power, seem as light as the carnival sledgehammer, the tool that proves you belong to a line of force extending backward from railroadmen to convicts to archaic pulverizers of rock and stone, men with those prehensile thumbs first used for seizing, then for making tools. You reach down into your pocket and pull up the money. At this bazaar, it's cash and carry, no credit cards for the Liquidators. We're paying off debt, not running it up. You grab the battery, the densest object in our dense display; we place the light bills in our cash boxes; and you carry off the battery like a player with a trophy. It's hand-to-hand business we do. No grocery baskets or shopping carts break the bond of new possession, evidence of the change in this exchange: the transfer of transport or the other way around, the transport of transfer, another almost anachronism—travelers meeting at a crossroads and conveying an object from far away, the burden of the distance traveled adding to the thing's measurable weight. We give ownership a caravan gravity now everywhere diminished by home delivery, UPS robbing the world of its essential weight.

One last secret, the Liquidators' final gift. We know you believe that so much stuff, like the lottery's huge number of combinations, must conceal something beyond our strongman's ability to bring the goods and our magician's skill in making them disappear: some buried treasure that's unknown to us, some thing like the million-dollar edition in the used-book stall or the $5 million painting in the junk shop, only more mysterious—gold secreted in a battery—or more miraculous—the bonanza lying just under all our feet. Surrounded by permutations of failure, you come to believe in long-shot success, not the self-helpers' "secret of success" that requires implementation and not just sudden wealth, but a find that will in a single instant define your earlier life as failed and the new life as fortunate, a word welded of luck and money.

Like tent shows rumored to heal those in failing health, we fail your secret dream, the personal surprise—one to a customer, one to only one of all our customers—that our show seems to warrant. What we give you is public, real. Leaving the arena, you notice the signs you didn't see above the doors when you rushed in: "Thank You for Contributing to the Liquidators' Savings." Strangely worded, our send-off recalls our invitation on the back doors of every trailer: "Follow Our Lead to the River of Savings." Having saved money in our flowing emporium, you leave as an immersed member of Midwest Liquidators. Holding your goods, you suddenly realize you're doing good. You too are giving aid, not full-fledged salvation of distressed businesses but the dignity-saving payment of some outstanding debts. Like us and with us, you're transforming total failure into partial success, participating in our fractional philanthropy and decimal deliverance. Satisfied customer, you pack your trunk or load your van, drive home the weight we've hauled across state lines. Lucky consumer, you're part of a purifying process as necessary as sewage treatment plants beside the polluted rivers of the Midwest we circle. Diffusing the collected waste of our nation's commerce, you're a local rep of the Liquidators' All-American altruism.

*II*

On the roof of the Middletown warehouse, where MIDWEST LIQUIDATORS used to sit before the company went out of business, is a sign that recycles most of the original letters: MUSEUM OF LEAD. You enter the front door at ground level, the only entrance. You buy your ticket—cash for individuals, credit cards for groups—and walk through a dark passageway to a stairwell. Down the winding, barely lighted steps you go, grasping the handrail, groping your way forward. On the uneven basement floor, cement as lumpy as the earth beneath it, you bend over or, if you're a large person, you crouch down to enter the tunnel before you, black as a grave, narrow as a grave. The floor is damp, the rough walls seep water, the irregular roof drips and is harsh on heads that refuse to bow. As you feel your way forward, a recorded message shrieks in a language no one understands and is followed by what you sense is a translation, an order: "Pick up your tools." You bend further over or drop to your knees and fumble on the damp floor searching for handles, familiar shapes and textures—plastic, steel, wood—but in this rock tunnel, all you find are pieces of sharp stone, possibly shaped but rough in your hand, stone age tools. As you stumble farther along, you hear stone hitting rock all around you, real or recorded you can't tell in the dark. You inhale what seems to be dust. The winding tunnel is not cool, as you expected. You're sweating now, twisting and turning and wondering when this will end, how long a day miners worked. The voice screams, "Drop your tools," and you do. "Move forward," and you do, anxious to be out of this hole in the ground.

You enter a room, close to six feet high, where a black man or a man in blackface sits on a stool. The man isn't dressed like any miners you've ever seen in photos. His clothes look like a gladiator's, and he's wearing crude sandals. In his right hand he holds an ugly-looking whip, multiple strands with little lead balls attached to the ends. He speaks gibberish to you, gestures to a wooden basket full of dark stone and earth, points to another tunnel. The person in front of you is entering the tunnel backward, hunching forward over the basket, smelling its contents, almost tasting them while dragging the basket into the tunnel. Your basket is heavy, and it smells bad, like the sensation when, as a child, you put your tongue on the metal of an unpainted swing set. You bump your butt against the walls because the tunnel winds and weaves like a subterranean brook. You are bumping backward into the past, forward, you hope, toward some light at the end of this tunnel.

After time you can't tell down here in the dark, you back out into a large room, dimly lighted. Another strangely dressed black man or man blackened by the dirt helps you heave the contents of your basket onto a slightly sloping table made of hacked wood. You stand and watch as water is released from somewhere above and courses through the piled earth, flushing away light matter, leaving behind heavier substances. With your bare hands, you and your fellow miners scoop up the wet remainder, load it into other baskets, and together carry them down yet another dark tunnel. You smell smoke; your eyes begin to water. You turn a corner and come before a scene like that in the haunted houses of your childhood: men dumping baskets into a contraption, perhaps a stone furnace, but you can't see clearly because of the smoke; other men, dressed only in loincloths, throw wood into a fire under the apparatus. All the men are coughing and hacking as they work; several men are lying on the ground convulsing. "Drop your basket, keep moving forward, ignore the slaves," a voice screams, so you don't have time to look carefully at the scene you've passed, ascertain if these were real men or only animations like those in amusement park tunnels of horror. In the smoke and shadows, the men doing forced labor looked like apparitions, and you wonder if mining and smelting preceded or followed the ancient underworlds you read about in school.

You enter a well-lighted room with white walls. From the ceiling is suspended a sign—"Lavrion, Greece, 1000 B.C."—and now that you've found your historical bearings, you understand the language and clothes and primitive technology you couldn't fathom in the dark. Under the sign on a large table is a scale model of this ancient mining village on the Aegean near Athens. Unlike the orderly towns of your model railroad youth, Lavrion is an ugly sidehill jumble of one-story houses, stone smelters, slag heaps, pathways, and water cisterns. Green is missing from this scene. The stone is gray, the earth brown, the trees lack leaves, and no grass grows in Lavrion. People are also missing, as if Lavrion were a necropolis or is now a ruins. On a plaque attached to the model is the truth of Lavrion's history: the silver in Lavrion's lead was largely responsible for the wealth of classical Athens. A voice says in a soothing mortuary tone, "Please exit to your right." But when you do, you're back in a dark tunnel and you think, "Hell is the place no one leaves."

You come into another well-lighted room. "Leadville, Colorado, 1890 A.D." is printed in large letters at the top of the wall on your left. Moving around the room, you see blown-up black-and-white photos of tents and shacks and pitiful people, strike-it-rich pioneers who look like Civil War prisoners. In the background are slag heaps, muddy roads, turbid brooks, clouds of smoke and dust. After comparing ancient ignorance and modern stupidity, you see next to the exit evidence of contemporary forgetfulness: a large color photo of Lavrion's harbor, blue sea and a few white sails surrounded on all sides by giant heaps of crushed black ore waiting to be loaded on ships. Next to it is a summer photo of Fremont Pass from the Leadville side: a whole mountainside gutted, white tailings flowing from the site. After this basement series of unpleasant surprises, you're pleased to see a lighted stairway, thankful that you're an upwardly mobile tourist in earth's bowels, that you've been saved from a past that's an eternity to those trapped in it.

At the top of the stairs, you enter familiar space—high ceilinged, windowed, and well lighted, a long and wide display space broken up by eight-foot-high room dividers. Embedded in the wooden floor is a yard-wide trough covered by plexiglass. In the trough moves a reddish liquid that looks molten. "Lead," you read on the first divider, "has always been valued for its low melting point." You follow the flow of lead as the trough zigs and zags a path that connects room-divided civilizations. In an area labeled "Mideast, 4000 B.C.," is a glass case containing dishes of a white powder meant to be applied to the face as a cosmetic and skin lightener. "White lead," you read, "was the first chemical industry in the history of man. Traces were found in the six-thousand-year-old ruins of Ur." In other cases are lead-glazed pottery in bright whites and yellows, fine necklaces and lumpy bracelets, lead amulets worn next to the skin to ward off disease. Egyptians brought lead to water: sinkers for fishing nets, plummets used to measure the Nile's depth, small lead statues of the liquidated and transformed god Osiris. Leaving the room, you read on a divider that "lead has always been valued as a casting material; some of the objects in this museum are true lead casts of original lead objects in museums around the world."

In an area labeled "Greece, 1000 B.C.," are cases with cosmetics like those used in the Mideast, many more samples of brightly glazed pottery, dull drinking vessels, candleholders, and a few imperfectly round coins. "When silver was scarce, Greeks made money from lead." They conquered other peoples using lead shot in slings and commemorated victories with toy soldiers made of lead. The Greeks cursed their enemies on lead tablets, taught children how to record history on easily scratched lead slates, and told the future by reading the way small drops of lead congealed in water. Lead was added to bronze to make heroic casts easier to produce. In one case are several "pigs" of lead, chunks and bars and wires used for caulking. "Lead," you read on a divider, "has always been valued for its abilities to contain and transport liquids."

In a large room labeled "Italy, 300 B.C., " huge photographs show Roman aqueducts built with lead pipe and repaired with lead solder. Cases display lead pots, kettles, tankards, and plates. Cooked and served in lead, foods were sweetened with a lead-laced concoction called sapa. Roman women continued to use lead cosmetics, began drinking wine treated with lead to retard fermentation, and applied lead to the cervix to prevent conception. To cure bodily ills, Romans went to baths lined with huge lead plates. Before their lead mines gave out, Romans were buried in lead coffins. "In the late Roman Empire," you read, "the use of lead achieved a per capita level almost equalling the level in industrial nations today."

The next area is "Europe, 300 A.D.–1750 A.D." Huge woodcuts picture German lead mines discovered in the late Middle Ages. You see the continuing use of lead to beautify, to seal containers of liquids, and to keep wine from fermenting. The English, with lead supplies of their own, produced a gray wealth of pewter plates, tankards, bowls, and spoons. Other display cases hold reproductions of old texts with magical drawings and secret designs, the pages dense with overlapping small pictures of odd objects, strange people, impossible animals, and alien symbols, circles within triangles within circles within rectangles. In one drawing entitled "Sol niger" a skeleton with angel to right and left stands on a black sun. Other drawings are violent: a swordsman poised over an egg, a seated king eating a child, a snake devouring itself. You learn the purpose, if not the meaning, of these designs when you turn a corner and find on a divider an eight-foot by twelve-foot blow-up of Pieter Brueghel's 1558 drawing *The Alchemist.* On the left sits a learned man reading a pile of texts and seemingly directing his two assistants—one in the middle puffing with a bellows and, on the far right, one mixing elements in front of a furnace. Between the assistants, a woman empties her purse while her children, ignored, play in a cupboard. The card beneath the drawing explains that alchemists first reduced matter before transforming it, and fine print on the card directs your attention to fine print in one of the alchemist's texts: "ALGHE MIST," Brueghel's Flemish for "all rubbish" or, more colloquially, "all shit." Laboring to discover the liquid secrets of wealth and health—the universal solvent that would transmute lead to gold and the elixir vitae that would prolong life—Brueghel's alchemist drives his family to ruin. Looking through a window in the upper left corner of the drawing, you see this same family in the future as, stripped of all possessions, they enter the poorhouse. In a curious touch, one of the children has a pot pulled down over his head, as if Brueghel saw ahead the long-term effects of heavy metals on the young.

A guide standing next to Brueghel's drawing ushers you and others into a tiny theater, perhaps twenty seats and a small, darkened stage that looks like a laboratory in an old horror film—alembics and torts and beakers. You can barely make out an old man with powdered white face and long white hair who sits motionless in a chair. He doesn't speak, but from behind him comes laughter and then words in a British accent: "Dead more than 250 years, and fools still pry into my papers and life, investigate my progress to gravity, speculate about why I went mad. Of course I performed alchemical experiments for twenty years. Alchemists knew force and sought the secret of matter I called inert. In 1693 I thought I would find that secret, but I failed and succumbed to distemper I blame on sleeping too close to the fire of my furnace. Just because I was afflicted by the gout, scientists ask to dig up my bones from Westminster Abbey. What laboratory would possess measuring devices that could prove that the bones of Isaac Newton, the greatest scientist of Europe, were permeated by lead?" The figure in the chair now begins to laugh, a low chuckle that rises to an insane howl. At the same time, smoke issues from the laboratory behind him, obscuring the stage and filtering out into the audience. In the dark and smoke that remind you of the basement, you hear the usher shout, "Exit left and go upstairs for the Industrial Revolution."

On the second floor, you're no longer being funneled and channeled, led forward through time by a trough in the floor. In the last 250 years of this millennium, you're free to wander among the display cases, free-standing exhibits, photographs, TV monitors, and flashing LED lights. The windows, you may read on the card beneath each, are leaded glass, sealing out the rain and cold. A huge display shows other protective uses of lead: in nineteenth-century gutters and downspouts, as a lining in drums holding corrosive liquids, in batteries filled with acid, as a sheathing on electric cables, in a wall for acoustic insulation, in the apron you've worn at your dentist's, in a mock-up of an atomic reactor. None of these would have been possible without the use of lead in printing presses, the machines that drove the revolution. You see hundreds of sets of type, raw data ready to be moved, printed, and transformed into new data. Another display shows a large array of lead weapons—irregularly shaped blunt objects, cannon balls of every size, homemade bullets and shot, then engineered hollow points and dum dums—the force that helped spread the revolution from Europe to and through the Americas. On one wall are photographs of huge pipe organs and stained glass windows in churches that have celebrated the revolution all around the world.

Suspended from a ceiling is one of those electronic signs that usually tell the time and temperature. On it flicks for a second some of the occupations exposed to lead while contributing labor force to the global revolution. The sign doesn't flash the obvious—plumbers, scrap workers, miners, smelters, the makers of lead flooring, lead salt, and lead stearate—but only workers who may not suspect lead is in their lives: auto mechanics, babbitters, bookbinders, bottle cap makers, brass founders, braziers, brush makers, cable splicers, canners, chippers, cutlery makers, demolition workers, diamond polishers, electronic device makers, electroplaters, emery wheel makers, enamelers, farmers, file cutters, flower makers (artificial), galvanizers, glass makers, gold refiners, gun barrel browners, incandescent lamp makers, japanners, jewelers, linoleum makers, lithographers, match makers, mirror silverers, musical instrument makers, patent leather makers, pearl makers (imitation), plastic workers, putty makers, riveters, roofers, rubber makers, shipbuilders, shoemakers, steel engravers, stereotypers, tannery workers, telephone repairers, temperers, textile makers, tinners, wallpaper printers, welders, and wire patenters. Below the sign in a mound of refuse are some mass-produced discards of the revolution—for late nineteenth-century women, "Ali Ahmed's Treasures of the Desert" skin cream, and for men, "Hall's Vegetable Sicilian Hair Renewer." More contemporary objects in the pile include empty wine bottles with lead foil, toothpaste and ointment tubes, tin cans with lead seals, broken-legged toy soldiers, cracked crystal, loose beads, divers' jackets, bottles that held moonshine liquor, and lead-glazed pottery—yes, still adulterated alcohol and lead-glazed pottery after six thousand years.

In the middle of all these displays is the most revolutionary use of lead: scientists' suspension of this easily liquefied but heavy substance in fluids. At the center of the floor is the liquidation of lead: stacked cans and drums of leaded paint from the nineteenth century, piled drums and barrels of leaded pesticides from the twentieth, and the centerpiece: a Stonehenge ring of old gas pumps—Esso, Sunoco, Amoco, Gulf—all the companies that moved Americans after 1925 when lead was put in gasoline. Suddenly a hologram figure of a traditionally dressed magician appears within the ring. When the figure stretches his arms, his black cape spreads behind him, making him look like the angel of death. "Before pulverizing the atom," he says in a voice that sounds like a ringmaster, "technology's closing act was dissolving lead in liquid and making that liquid disappear into thin air, turning the air thick, polluting our cities, our farms and forests, even the polar ice caps thousands of miles away" To the sound of an engine slowly revving, the magician dissolves into a holographic black cloud that seems to rise and spread beyond the ring of gas pumps.

You wonder about this museum and your progress from earth to air and back to earth. Museums display rare achievements, not widespread failures. Even exhibitions that aren't really museums show labor-saving devices and life-saving objects, not forces of destruction. You brought your children here to view things, not to be addressed by frightening voices. But wander as you may through these exhibits and performances, you can't exit backward. Failure lasts; failure continues. This is, you now understand, a forced tour, and the only way out is upward to the third floor, forward to the last words.

The first thing you see is an ancient Greek sarcophogus made of lead to hasten the process of decomposition, introduction to this story's effects of lead. On the wall before you are contemporary artists' renderings of Egyptian and Greek faces, eyes destroyed by lead mascara, cheeks pocked by lead rouge. Around a divider is an enclosed craftsman's shop, the artisan hunched coughing over his leadwork, a curse in Greek letters. Farther on, TV monitors run clips from old films about the Fall of Rome, deranged emperors who were, you read in subtitles, "besotted by lead-heavy wine and, we now know, suffered from gout. Rulers and aristocrats did not reproduce. The line failed." Then you see enlarged medieval woodcuts of women and men holding their stomachs, poisoned by adulterated wine and cider. Old engravings show well-dressed European and American men made motionless by gout. There are charcoal drawings of nineteenth-century lead factories surrounded by wastelands, vegetation dying, livestock dead. You stop to read about poisoned workers, an enlarged page from Dickens's "Uncommercial Traveller." A photograph of dead swans is accompanied by a newspaper headline: "Queen's Swans Killed by Thames Sinkers." A huge color graphic shows gray pesticides leaching into blue aquifers in the United States. Another wall-sized graph shows rising parallel lines from 1925 through 1975: the use of leaded gas in America and lead pollution. "At its peak," you read, "leaded gas contributed two-thirds of environmental lead contamination in North America."

You turn a corner and enter intimate reality, an inner-city apartment, this year's calendar on the refrigerator. Paint is peeling from the walls, and chips are on the floor along the baseboards and under the radiators. No adults are working or resting, no children are playing, but a recorded voice instructs you in weight, small amounts, heavy micrograms: "A child who licks a finger dusted by this lead paint doubles the maximum daily absorption of 100 micrograms." Beyond the apartment, there are life-size models of a man and woman, transparent plastic skin over working organs. They harmonize in synthesized voices: "From air and food and drink, lead enters our systems and is spread by the blood. The toxin affects the stomach, kidneys, and brain, the organs with which we process substances, liquids, and information." Now the voices begin to slow down and break apart. "Anemia, gastritis, nephritis, and peripheral paralysis are the continuous diseases in adults with continuous exposure." The wrists of the models drop, go limp. The voices slow further. "Retardation and brain damage are permanent effects in children." You turn another corner and see a lead-lined coffin, advertised to give "the loved one permanent protection from the incursion of liquids."

But the history of lead is not all failure. Moving toward the exit, you see graphs showing the steep reduction of lead in North America since it was removed from gasoline. You see before and after pictures of smog-bound American cities, as well as photographs of cities in countries that still use leaded gas. For those with occupational exposure, contemporary medicine everywhere has the solution in chelation, a washing of the blood like the washing of liquid waste. Brain damage can't be reversed but can be prevented with aggressive cleanup programs in the decaying neighborhoods where lead can get into children's blood and bones. The very properties that led to lead's diffusion make it recoverable, recyclable, melted back to a liquid state, and reformed in solids we are now certain are dangerous. Humankind is not saved yet from long exposure to what lasts, but leaving the last room of this museum you sense success. It's a partial success, you have to realize, because in the future, scientists may discover that lead is even more pervasive than they now think or that present threshold levels are too high. Partial, too, because lead is but one of many heavy metals, metal is but one of many pollutants. And pollution is but one of many failures in the past that have run up future debts to the earth. Success will be gradual, recovery slow like the decay of lead.

You're out of the building now, standing on the open stairway at the rear of the Museum of Lead. It's been a long tour through the four stories of information, and you hold no souvenirs. Instead of the power you felt exiting the Liquidators' show, you may feel chastened and small. And yet, looking down at the Great Miami River, you might also be surprised to feel pride in our million-year-old ancestors. Humans settled near water—springs and brooks and rivers and lakes and seas—and learned its force: how it sprang from the earth, moved at its will, tumbled stones, covered land, and made things disappear. Slow learners, we were, picking up the minerals that came to hand, boiling water to break up rock, using stones to hammer stones, but courageous too, burrowing into the earth, making our own caves and caverns like underground rivers, slowly understanding how liquids could be used to separate substances, how force could be employed to pulverize substances, how fire could purify substances and transform them into liquids we could form more precisely and repetitively than carved wood or chipped marble. We didn't just hitch a ride on water or divert it for irrigation or channel it for power. We tool-makers used our force and ingenuity to liquidate substances, make weapons to protect our offspring, make money to pass on to them, make life longer to project our genes. Like lead, we spread and dispersed, a global tour de force, digging and pulverizing and liquidating and reforming the world like malleable lead. We made weight and the machines to move it. We cast pipe and brought water into our homes. The Ages of Man since Stone are Liquidator Eras: brick, lead, copper, bronze, iron, steel, glass, oil, plastic, all liquids become solids in our service.

We're all liquidators. But much as we might wish to celebrate ourselves, we must recognize that our magic power became an obsession and that we began liquidating ourselves. Now we've inherited substance abuse, our centuries-long addiction to transforming substance into liquids, fluids, and gases. Still, no matter how dense and dull we have become we can reform, and like the Great Miami moving slowly in the man-made channel it will never overflow, we can adapt to limitations we cannot reverse. Yes, we're all liquidators and must resolve together, become a global band of self-serving altruists.

Midwest Liquidators was wrong. The bottom lines are these. Not everything flows or fails. Something surprises. The Liquidators no longer barge into your town or city, but cloudbursts, flash floods, and tidal waves remain. Although you never found the secret treasure in our show, surprise is certain still. Like lead lasts, surprise is sure. When we failed, nobody predicted the Museum of Lead in Middletown, Ohio. Full of earth's weight, the Museum of Lead and Liquidating is finally light, uplifting like the spurt and spew of geysers, those spirit plumes forced skyward by the earth's liquid core.

## Dragging the River

*Sean Gillihan*

Vernon Dombrowsky:
Eight days already and things don't look real good,
but it's not like we haven't been trying. You said it,
it's the needle in the haystack routine.
Things are so different underwater,
up near the face of the dam
it's nothing but holes and roots,
any of them big enough
to hide a body or, in this case, two.
But the bottom is changing every minute,
we're so close to that dam.
You have to be careful.
You want to blame these kids, focus on something,
just so your mind keeps running, stays warm.
You play it over in your head
again and again,
you put yourself out in the boat
right when it goes over.
Sometimes I can feel my lungs working
then filling, see the way light scatters below the surface.
Think about sound underwater, how it's just pressure really.
But this is no one's fault, nothing we can do
about it now. I've seen it before.
I tell you the coffee couldn't be hot enough
on a day like this, the cold goes so deep.
It could be weeks or more, a lot of time
it's just waiting.

Adam David Clayman

# Tainted Torrent

*Robert Grudin*

### *The Womuba Apocalypse*

Monday, April 6, 1998

I have walked in the glory and felt the horror, I have snatched a glimpse of the promised land and slid back into the pit of Hades, I have tasted joy's grape and dashed it into smithereens against the wall of folly. In short, I've fucked up.

I've had it all and lost it all—Mara, the money, the van.

I'm in a cell in the county jail, Pilotsburg, KY. It's grim and fluorescent and overheated in here, and the overheat is vaporizing the disinfectant in my uniform and making me smell like a latrine, and in fact I'm sitting on the toilet because there are no chairs and the bed's too soft, writing on a gray confession pad given me by Schultzie, the sheriff's deputy.

Not that I'm minded to make a confession. These notes will become a confidential letter that I mail out of here, which is part of my rights in this jurisdiction.

My stomach's sort of bloated and acidy, from having drunk a portion of the Sacannah River when the water wasn't so clean.

Schultzie's a red-faced sausage of a man who seems to be okay. He's taken a liking to me, maybe because, as he puts it, it's not every day you see a big man cry, and I'm crying now, even as I write, forgive me World, for what I've lost.

It's still raining hard outside, the same apocalyptic demon storm that screwed me up yesterday, and everybody here in the slammer is talking flood. The Pilotsburg levy is twenty-seven feet, and they expect the Sacannah to crest at thirty-three feet around three this P.M. What happens to us prisoners? Only the Sheriff knows, in his wisdom, but right now he's up in Louisville, defending himself against a construction fraud indictment, and he's cellularizing Schultzie at short intervals.

It's okay boys, leave me here to drown. I'm grossed out on life.

Blame the entire mess on the Vixen. Here we are, yesterday morning, driving south because I decided (shit!) that getting on I-40 would buy us some better weather. As I drive I'm drinking lukewarm instant coffee, while Mara regales me about how her aunt and guardian, Big Cuerva, owned three pets that she kept right in the house—a horse, a goat, and a parrot. I'm starting to giggle about the parrot, who liked to quote bawdy tidbits from the *Song of Solomon,* when all of a sudden the rig loses power, and there's no way to pull over to the right because I'm in the passing lane in traffic. I pull left onto a little grassy median that gives me about a yard of clearance from the fast-lane traffic, and as luck would have it, the engine's to the rear on that side. Wait for help? Fuck that, I don't want to schmooze with any state troopers. Letting the engine idle in neutral, I grab my tool kit and slide out the driver's door and, feeling doomed in the strange warm air, am crouching and opening the engine compartment as God knows whatall races and roars just behind me. In spite of all this excitement, there's still room for a nasty shock when I see the engine. A BMW diesel!! Beemers are great but quirky, and for motor-home use I'd have preferred something much plainer, like a GMC. As traffic fans up warm air around my buttocks, I lean deeper and squint.

Sure enough, this engine has quirked out real good. The throttle works fine, but the throttle line's corroded and parted right at the carburetor take-up, and there's no way of fixing it out here. But what's that rattling? A double whammy: the clutch in the fan is shot!

So what's the problem? In my current mood, these things don't bother me at all, for somewhere along the road to insanity I have developed Self-Esteem. I inch around the back and sit down in the driver's seat, thinking hard.

"Tell me the matter!" says Mara, wide-eyed.

I tell her.

"Can you fix it?"

"I think so." She gives me a long look, like I'm some kind of magician. I tell her to go back to the duffle bags and pull the tie-cords off them and to tie these cords together at the ends to make one big line. When she reappears I test the connections: I've now got a thin, strong line about twenty-five feet long. I turn off the ignition and tell her to start it and rev it up again when I bang on the side of the Vixen. I bend over Mara and throw most of the line out her window, leaving her with one end, and in short order I'm around the front taking up the line. Back at the engine I disconnect and remove the fan belt and tie the line to the throttle. Then I bang on the side of the truck. Engine starts, and I yell at Mara to pull on the line.

*Vroom, Vroom.*

I get back into the cab. Mara looks stricken when she sees the fan belt. "You mean it'll run without that?"

"I aim to find out."

We're truckin' again, sort of. Through trial and error I learn how to pull or ease up on the throttle line as I go up through the gears. This works okay, it turns out, but the situation forces a change in plans. We've got to make Louisville, which I'd hoped to bypass, so as to do repairs, and we've got to make it by the levelest possible road, because I'm gonna have to remount the fan belt before we climb every hill. On an easy stretch, where I don't have to shift, I take the throttle line and tell Mara to open the road atlas to Kentucky and find a red line to Louisville that looks straight.

It turns out the next exit was as good as any. Or so we thought.

Fate takes strange ways. Immediately off the highway we were simultaneously hit by the smell of what seemed like (and was) 10,000 barbecuing chickens and consumed by the desire to eat a couple of them apiece. I guess that's what emergencies are always like—you go through them in a kind of trance, and when they're over, you all of a sudden need food. The irresistible smell led us a quarter mile upwind to a pair of huge gates opening on a humungous grassy lot where every vehicle in Kentucky seemed to be parked. Between the gates was a billboard with permanent brass letters announcing that this was the Womuba Recreation Area and big movable letters proclaiming that today was the

5th Annual
United Charities Chicken Orgy

and that the entertainment would be

PLATINUM JAYNE
AND THE GROAN

Mara and I looked at each other in surprise. Charity chicken orgies aside, what really boxed us was how they'd gotten Platinum Jayne and her Groanykins to come to Kentucky, which was more or less like getting the mountain to come to Mohammed. Last time I'd seen Jayne on network news, she was doing a benefit wingding with Madonna and Pavarotti, who'd both come (with about 65,000 other folks) to her entertainment palace in Aspen. Word was that the White House liked her for the Medal of Freedom but was afraid she would come to the ceremony in costume (or lack of same) and make Daughtergate look like so much spilled pablum. Platinum Jayne! Even the demure Mara whispered the name in amazement.

At this point we were slowly weaving our way through the unpaved parking lot between cars and vans and tailgate parties that involved every social class, age group, and lifestyle, all clumped together in a Sunday mood of group frenzy. Even from the cab of the Vixen, you could smell a summery medley of booze, sweat, and pot. I was then about—oh God!—to take a turn into a vacant parking place when Mara grabbed the wheel and shouted "STOP! DON'T MOVE!" and was out of the van like a shot. In seconds she was at the driver's window holding up an open, empty-looking Budweiser case that I'd just been about to run over.

It had a live baby in it. The baby in the beer case was smiling up at me.

I parked the van and found Mara nearby, kneeling in the grass next to a motionless supine woman whose pregnancy looked as though it'd run beyond term. Passed out on beer, indicated empty cans. Behind her, a red-headed man in a terrycloth robe emerged from a beat-up microbus and took delivery of beer case baby with no word of thanks and a holier-than-thou look.

Feeling faint, I remembered having left the keys in the Vixen. Mara went back to get them and lock up. And then, together, we walked out through the lot and, toward chicken aroma and distant music, onto a graded road that led us up a little rise. The air was uncannily warm, like around 75 degrees, and we were walking in sunlight toward black clouds hiding behind that rise. Blossoming cherry trees lined the road and blackbirds were singing in them, and I remember feeling for a second as though we were Adam and Eve (not that I was in shape to have carnal knowledge!) going forth into an unknown world.

"How'd you know the baby was in that case?" I asked.

"*I had a feeling,*" she whispered.

"Thank God!" I burst out. Was there something extrasensory about this girl? Of course not, I thought, looking down at her as we walked. But the look of recognition that those aqua eyes shot back at me made me think she'd known what I was thinking at that very moment!

Just then we topped the rise, and a scene unfolded that I'll never forget.

We looked down on a vast panorama.

Imagine a flat plain about three-eighths of a mile square, stretching from the rise and dropping off in cliffs over the Sacannah River. Imagine the whole left-hand side of this plain being full of people, all of them facing a big stage that was right at the edge of the cliff. Imagine on the right smoke rising as thousands of chickens are barbecued in huge pits. Imagine all this lit by a sun that's slowly being smothered by mammoth black clouds, throwing off a light of supercharged lurid whiteness, like concentrated moonlight, without a touch of gold.

Now imagine a wall of noise, heavy metal to the tune of 200,000 watts, almost drowning out the even deeper noise of approaching thunder.

The air must have been very heavy because the noise was drubbing us almost like a wind, louder and louder, and the crowd was so big and scattered, with people straggling in and out all the time, that we could get almost to the front row and see everything. What a sight to behold! We were maybe 35 feet away from the opera-sized stage, and standing on this stage to the right was a very colorful but mysterious group of people. Each of them was holding onto a lifeline attached to a big orange balloon that floated in the air above him or her. On the left hand side of the stage, the Groan were making big-league noise in their famous rendition of "Sixteen Tons," with Fallic Devine sitting (as usual) at the drums in front of everybody, and Ditto Kelly and Jerusalem Scratch (just back from rehab) pumping guitars right behind him, and tall Orville Negombo bouncing up and down in back as he punished his bass boomer, so that from the front they looked like one superhuman creature with lots of never-stopping arms and legs.

But nobody was looking at the Groan. How could they, with Platinum Jayne standing center-stage? Not that she was singing or dancing—she was just sort of undulating to the music, but what undulating! It was as if she was the conduit the music was flowing through. As always she was topless, and as her arms and torso moved, her boobs followed in private epicycles, as though they had minds of their own. And the metal! Believe me, Jayne was fitted out like a steamship! Three rings in each nipple for starters, a ring through the navel, and as she turned and you could see the ass-cleavage under her Daisy Mae skirt, a large ring through each buttock. Down her left inner thigh ran a chain with a locket attached to it, also wigglewaggling to the music and leaving its point of origin to the imagination. Bracelet-sized earrings, elbow rings, knee rings and smaller but shinier equipage on the eyelids, nose, lips, fingers, and toes.

I was still counting rings when the music stopped, punctuated by a heavy crack of thunder from behind us. I looked down at Mara, who looked up at me with a frown, suggesting maybe that she hadn't liked the music, or maybe that it was going to storm.

*That was the last time I saw her.*

Platinum Jayne's voice was suddenly drowning out everything. Standing tall at the mike in the deepening gloom, lit intermittently by lightning from the approaching storm, Jayne launched into a major speech. First off, she explained how this whole performance was aimed at the welfare of disabled Americans everywhere. Next, she introduced the folks with the orange balloons—Flo, Fuzzy, Herschel, Baxter, Pedro, Wendell, and I forget the rest—as disabled Americans from twelve different states. They're standing up, she said, for the first time in years, thanks to their being fitted with body braces attached to helium balloons, and even those whose legs were totally paralyzed could get around, in a vague sort of way, with the help of special canes (as an example of which Fuzzy, the oldest of them, navigated his way up to the mike amid applause). Jayne allowed as the technology of this balloon stuff wasn't very well developed yet (*she was right about that, poor devils!*), but that she'd bankrolled these getups so the Womuba Twelve, as she put it, could be symbols of the quote-unquote Empowered Body. Then she introduced Fuzzy, a smallish bearded coot who looked as though he'd rather be somewhere else, like maybe at the dentist, as Ira "Fuzzy" Blomsky, her own father.

The rest of the Womuba Twelve didn't look so happy either. They were all looking helplessly westward at the approaching storm, and for an instant I saw lightning reflected in their eyes.

But none of this fazed Platinum Jayne. With breasts heaving and rings clinking, she was now into the main course of her performance, a singing sermon that she called "The Gospel of Bod." Her sermon really rattled me. What did she think—that she was going to convince this huge audience, a lot of whom were probably Baptists, to turn away from Jesus and start worshipping the human body? Many in the audience were getting fairly riled up. I looked back to see a number of pretty tough-looking customers standing up and shaking their fists. But lots of others were cheering. Accompanied by the Groan, Jayne started belting out this medley of hymns, with *Bod* always substituted for *God,* as in "Nearer, my Bod, to thee," and "A mighty fortress is our Bod."

*Then came this colossal roar.* I looked back at the multitude, thinking for sure that this was a riot, but it was worse. It was some kind of devil wind, come up out of the river gorge, and it blasted down on the people behind, so awesome and loud you'd have thought it was some colossal attack helicopter right out of *Apocalypse Now.* A tornado, for sure! Like a big angry dog, it tore into the roasting pits behind the seats, and in a moment all of us were being barraged by a hail of flying barbecued chickens, whizzing sizzling featherless birds, and I'm afraid some people may have been badly singed.

But even while fending off chickens with one hand, I couldn't keep my eyes off Platinum Jayne. In the melee she stood firm, grasping the mike and shrieking something about Bod with all her might, and in the middle of that shriek, I shit you not, the sky opened and a bolt of lightning as thick as a tree transfixed her as she stood. There was a thunderclap such as almost knocked me down, while the lightning, which seemed to hold itself still in time, played horribly over all her rings and ornaments and her voice stopped and Platinum Jayne burned to a crisp before my very eyes, an ashen crust that was quickly swept up by the wind.

Then in a flash, as luck would have it, things went from bad to worse. The wind had died down for a minute, and I'd barely realized that the smell in the air was Platinum Jayne's roasted flesh—which was especially gross because it had all the chemicals from her melted rings mixed in with it—when I noticed that something was very wrong onstage. The Womuba Twelve were taking off into midair! It must have been a sudden drop in air pressure—or was it a sudden rise?—but there they all were hanging from their orange balloons, floating five, seven feet in the air, arms wagging helplessly about as mischief-gusts whirled them in little circles around each other, some shouting "God! God!" and others "Bod! Bod!"

I had to help them. I bounded forward and jumped up onto the stage. Old Fuzzy was in the greatest danger; some kind of felon wind had caught him alone and was pushing him out over the gorge. I hate heights, but there was something about Fuzzy's utter helplessness that drove me on. As I made for him, he'd floated just past the back end of the stage, which, as I immediately found out, was the edge of the cliff itself. He had nothing under him but 300 feet of space to the rocks below, and the cliffs on the other side loomed gray behind him, and he had this sad, confused, wistful look on his face, and as I reached out to him he asked, "Are you Bod?"

That's when I fell off the cliff.

In midair I grabbed Fuzzy and held on for dear life as wind rushed up by us faster and faster. My face was all fouled up in his beard, but I knew that, balloon or no balloon, we were headed toward the rocks at increasing speed. I reached out an arm to change our trajectory, but it was no use, and we both would have been gull feed if it hadn't been for a terrific river updraft—the same sort that must've scattered the chickens—that cushioned us and sent us careening off toward midstream. The Sacannah is about half a mile wide at that point, and we made splashdown smack in the middle.

The first thing I remember is that the water had a disgusting warmth to it, like falling into blood, but worse than that it was cloudy and foul. I could taste all kinds of flood crud in it, like car and animal and house. For a few seconds I couldn't find Fuzzy among the brown whitecaps. But then I heard his cry above the din of rampaging water. I paddled over and grabbed him. I'm a good swimmer, but I was being dragged down by my shoes and clothes, and the balloon was now partly deflated, and it was all I could do to keep the old guy afloat. I could see him swallowing water, and I couldn't do jack shit about it. It was raining buckets now, making things even harder to see, but I made out the large form next to us, and whatever it was had a leg, so I grabbed the hock. Turned out to be a big dead bloated pig. Holding Fuzzy with one hand, I held onto the hock with the other, for flotation and balance and the stability of its mass. Now as a summer rafter I know a little bit about rapids, and I guess that's what saved us. I knew that the water would be faster in the gorge, because gorges are steeper, so the gorge was no place for trying to get out of the river. Beyond the gorge, things might slow down, and with luck I could let go of the carcass and paddle toward the inside of a bend, where water tends to eddy. Even then I would need strength and luck pulling Fuzzy onto dry land, not to mention what it would take to revive him and get him to a medic.

Maybe from the tension or who knows what, I was getting groggy, and I must've zigged when I should've zagged, because I was dragged across something sharp, maybe a snag, that opened up a few inches of my right arm. I was bleeding into the river now and slightly zoning out about the evil results accruing to all those who celebrate the Gospel of Bod, when an image flashed into my mind's eye, complete in detail and clear as day. It was of a walkway back at the University of Oregon, where the trail passes between the School of Ed and the backyard of AXΩ and a willow tree from AXΩ hangs out over the path. Spring of junior year, just after I had to quit basketball when I hurt my back in that collision with Shakeem, I would meet Glynda there, days when she got out of her Ed Psych class and nights for secret kisses. She'd rub my back where it hurt and try to comfort me. That spot behind the frat house meant happiness to me, memory as sweet as dreams. I was trying to open my eyes—you've really got to keep on your toes at times like this—when another image zapped me. I was a little boy, maybe six years old, walking with Uncle Titus (who looked old even then) in Golden Gate Park in San Francisco. After much pleading and whining I'd just persuaded him to buy me an ice cream cone—the kind of vanilla they don't make anymore, that had little pieces of real ice in it, and the ice cream man mustn't have put the scoop on very tight, because as soon as we left the stand, the cone tilted a little and the ice cream fell right off and plopped onto the sidewalk. I looked down at my lost joy, and I looked up at Titus's accusing eyes, and I wanted to disappear from the face of the earth.

Well *that* woke me up all right. I opened my eyes at the moment we were rounding a right-hand bend and I saw what I'd been waiting for, a village on the inside bank. It had everything a drowning man might want—docks, ladders, and ramps—but it was almost too damn close, and I dropped the pig and paddled like hell with my right arm. It was a nightmare watching that village go by. Even in the rain I was close enough to see the faces of those who reached out to help me, a red-faced clergyman (I got close enough to smell the booze on his breath!) and, next to him, a strikingly beautiful black woman in a red mackinaw who almost fell in trying to reach us, and finally, at the last put-in, a kid wearing an Atlanta Braves cap and holding out an oar. As my fingers slipped from the oar, he shouted something about a mill. I went under, swallowed about a gallon of river, and came up passing a bank of willows. Sure enough, there was the mill close ahead, an ugly old monster with a white stack and, just for Fuzzy and me, a mammoth intake port that must have led down to a nest of hydraulics. And damned if I didn't see a big paddle-drive down there, waving to us, waiting to turn us into gruel. *Not on your fucking life!* I reached up and grabbed a willow branch and begged it not to break. The branch bent around until Fuzzy and I had completely reversed positions and I was hopelessly strung out between him and the limb, but, thank God, it dumped us in an eddy.

I let go of the branch, found my footing and—strange that my back didn't hurt me—pulled Fuzzy onto the bank. He was unconscious. I undid his body brace, pulled him away from the brace and balloon, and did what resuscitation I could. I was still at it, with no positive results, when a pickup pulled up near by. Without really meaning to, I lay down on my back. Before I passed out, I recognized the Reverend and the boy in the Braves cap, and above them, finally free of its tether, the half-inflated orange balloon wobbling up through the willows on its way to heaven.

### *Dreamer, Awake!*

I must have been unconscious, or semi-conscious, for eighteen hours. You know how sometimes you'll be lying in bed, late at night, thinking of something grim, it seems like forever, and then look at your watch and see that only a few minutes have passed? In those eighteen hours I dreamed my way through a lifetime of loss and loneliness.

A hundred times I was with Mara, carrying her, helping her, chatting with her, feeling the warmth of her presence, or I was back years ago in her childhood with her, eavesdropping on her as she sat on Aunt Cuerva's knee.

A hundred times I fell off the cliff and into the foul river, swept past all hope of return.

The next morning—this morning—I woke up in this cell. I'd been washed down and dressed in jail clothes. My right arm was bandaged and my other arm ached, probably from a tetanus shot. When I could stand up I went to the john and pissed out a tributary of the Sacannah River and went to the door and shouted. This bought me a visit from the Sheriff himself, big hawk-nosed honcho, chewing a foul dead cigar, introducing himself as Luke Thrumble, reading me my rights in a sadistic drawl, and telling me that I was going to be charged with the murder or attempted murder and/or kidnapping of Platinum Jayne's father, Ira "Fuzzy" Blomsky, but that my case wouldn't be "processed" until (a) the flood danger was over and (b) Blomsky had died or pulled through—they weren't sure which—and they could then decide what was the more damnable charge they could pin on me.

Murder! Ira Blomsky! My selfless act of mercy mistaken for a crime! I was so eaten up with shock and rage that I almost keeled over on the spot, which gave the Sheriff his chance to get out of my cell and hightail it to Louisville. I lay down on my cot and thought unpleasant thoughts. About how hard it would be to prove my innocence. About how, even if I proved it, investigators would connect me with the Vixen and the money. About the freedom I'd had before the Vixen broke down and we got hungry for chicken. I started crying, but left off when Schultzie appeared with a tray of biscuits and gravy and some black coffee. I was wolfing and slurping down same when Schultzie came back with the Reverend, who was shaking rainwater off his umbrella. The Rev, a fine figure of a man with a handsome red nose and a blue-eyed look at once suspicious and comic, as though he was trying to sniff out a joke, introduces self as Howard Keane, sits himself down, and, apparently aware of my depressed state, offers me a slug of bourbon from the biggest hip flask I've ever seen. Over this profoundly refreshing drink, which he said was made by a family named Wharton, I told him my plight, skipping of course what brought me to Womuba Park. He'd already seen on TV about Jayne's tragic end and how a stranger named Bod had made off with her dad, but he was especially interested in my end of it, and after each new little episode of the story, he'd nip on his flask and shout, "Capital!"

"What d'you mean, capital?" I finally asked.

"Mr. Bod," he said, "you are a historian of the times." I couldn't get him to explain this, but he was off on a tangent, some of which made me laugh in spite of himself, about people who'd been wrongfully accused or otherwise had gotten ear-deep in shit. He ended up by saying, yes, that he knew of eight or nine people who were even more unfortunate than I, but that they were all dead. All, that is, except one.

"Who's that?"

"Mordecai MacCrae," said the Rev.

"What happened to him?"

Priming himself with sour mash, Rev described MacCrae as an old Gulf War buddy, a war hero and "the most capable man I've ever met," later an L.A. cop whose exploits had been (though he kept this a secret) the source of three hard-hitting Hollywood movies. MacCrae was brought to Louisville on special assignment, because the Feds wanted a good man who wasn't known in the area. Peter Carrarra, who was a big deal in the FBI (I'd heard of him myself), and Taylor Munson, the local DA, wanted an informant on Death Row at the Kentucky Pen. They wanted the goods on Eddie Moon, the underworld kingpin, and figured they could get them from Eddie's pal Fats Vukovich, a Death Row inmate. But the plant had to look legit. Carrarra and Munson met with MacCrae in total secrecy and asked him if he would take a bum rap (under an alias) for rape and murder. He agreed, as that was probably the only way to get Moon. So the next time they had a suspectless crime, they pinned it on MacCrae, and he was cycled through the system to Death Row. And there he is, getting the goods on Eddie Moon, when he sees in the paper that Carrarra and Munson have both been gunned down, in two different cities, *on the same day,* and that they were both deader than King Tut. Seems that Moon had a contract out on 'em. And Carrarra and Munson *were the only two in on the secret.*

"How come you know it?"

The Rev said he'd visited MacCrae in prison.

"Why can't you help him?"

Because he'd only MacCrae's word to go on.

"What happened to MacCrae?"

He's still on Death Row. The stays and appeals have almost run out, and he's got about nine months to live.

Suddenly feeling ornery, I told the Rev that if he knew my full story, he'd see that I was even unluckier than MacCrae, because I had more to lose.

The Rev looked up in mock surprise. "MacCrae's a good man. You don't have more to lose than that."

That stung. A good man? I hadn't heard that phrase in years. And whatever it meant, it pretty well didn't mean *me.* For a moment I wanted to stop the Rev and ask him what exactly made his friend MacCrae such a good man, and what makes a good man in general. For a moment I forgot my money and my danger and my own miserable selfish lot.

But the preacher was still talking. Hardly pausing, he made a broad sermonizing sweep with his right hand. "Now your Sheriff Luke Thrumble here, he's got an awful lot to lose, but he's not such a good man. He and his brother Art (who's a real asshole) own most of this old town. But if they don't watch out, the state's attorneys are gonna take it all away—that is, if there's anything left of Pilotsburg to take away."

I asked why.

"'Pears a while back, the Sheriff and brother Art put in the low bid for building the Sacannah Dam, and now it transpires that instead of building it with rebar or concrete or whatever expensive stuff you're supposed to build dams with, they dumped in 100,000 tons of landfill, and nobody found out about it till the County went to repair a hole in it, and out pops an old washing machine."

I asked if the dam in question was upstream of us. "You bet your balls it is!" exclaimed the Rev.

"Then how come the river was flooding when I fell in it?"

"If you call that a flood, you ain't seen nothin' yet. If that dam breaks, you're talkin' wall of water," said the Rev, refreshing himself with more bourbon. His face was beet red now. Before my eyes, in the middle of the day, this preacher was getting pie-eyed drunk. I wondered how this could be. Was something gnawing at his guts?

I wouldn't have found out if I hadn't asked him about that beautiful black lady by the river. The poor guy starts sobbing! Turns out he's head over heels in love with her! But there's more. The Rev is gay, or at least thought he was. All this goes way back. Until the Gulf War, he was celibate, though he didn't have to be, being Protestant. But as a chaplain under fire he began to feel love for his comrades. It was a kind of love he couldn't understand, seeing as in the church and the army homosexuality was swept under the rug, so to speak—I mean, not discussed. He thought there was something wrong with him and tried to cover it up. But it got to be too much for him, and one day he more or less propositioned the company captain, Mordecai MacCrae.

MacCrae turned him down politely, but told him he didn't think there was anything wrong about being gay. All that was wrong, he said, was to have emotions and keep them a secret or lie about them.

But the Rev didn't see things that way. "God's work comes first," he'd told MacCrae, and he knew that he wouldn't be allowed to do God's work as a self-confessed gay. Back in the States after the war, he set up shop in the church at

Pilotsburg, and for years he cooped up a secret he couldn't tell and passions he couldn't express. Imagine that!

The Rev got resigned to this fate, and even began thinking that it was God's way of punishing him for his lust, until a couple of years ago, when he met Heidi Frederick. She's a social worker assigned to his church. Right off he was amazed at how she worked with the poor and the suffering. "She was getting paid for it, but you wouldn't know it," he said. She worked as though helping people came naturally, even as though it was some high-class kind of self-expression. It was the first time he'd ever respected a woman professionally, and what made it worse was that in her case, professionally meant spiritually. Add to this that she was obviously some kind of genius in understanding people's problems, and you've got a powerfully attractive mix.

And her manner with him! Direct, warm, and without a hint of female-male or black-white uneasiness. Everything she said or did was full of heart.

One morning last summer Heidi and the Rev ran into each other under an apple tree in the parish garden, and he realized he was in love with her. (I know exactly how this is, because it happened like that to me with Glynda.) It was very warm, and she'd left an extra buttonhole of her blouse unbuttoned, and when he'd seen the top curve of her breast and smelled her morning soap on her, he was suddenly a raving hetero and hopelessly in love. *How do you explain these things?*

The Rev said nothing about this to anybody, especially Heidi, because he had this habit of stewing over things secretly rather than blurting them out. He says those were the most precious weeks of his whole life, the weeks when he worked with her, chatted with her, all the while silently adoring her every move and word.

I asked if it mattered that they were of different races.

Yes, he said, the race thing made his love and desire that much more intense. He saw God in her, but it was like no God he'd ever known before—it was a God who coursed through his veins like fire. In the end he couldn't hold out. After all, why should he? One October evening he found a way of getting her under that same apple tree again, the apples being ripe this time, and he told her that he'd loved her for many weeks.

"What happened then?" I asked.

"Then God struck me down," said the Rev. Heidi replied that she had lots of liking and respect and even love for the Rev, *but that some years before she'd discovered she was a lesbian!*

What irony! How do you sort out a mess like that?

"You don't," said the Rev. "Except maybe you pray." Saying this he staggered across the cell and took a long loud piss in the john and then went and banged on the bars to be let out.

I asked him if getting drunk was his way of praying, and he half turned toward me and said it was his way of staying alive so he could pray.

As the cell door slammed shut, I fell back on the cot. I must have been asleep before my head hit the pillow. I slept about an hour and had a single dream. There was Heidi Frederick by the river again, reaching out to help a drowning man, her face so beautiful that you didn't know whether to kiss it or pray to it; but the drowning man wasn't me or the Rev. *It was my father!* I yelled "Hey Dad, watcha doing there?" (how's that for a silly question?), and instead of answering he gave me, from the waves no less, one of those long, sad looks of his, those sort of Oliver Hardy looks that suggested he knew some fault or limitation of mine that was such a bad one that I'd never understand it myself no matter how hard I tried, and at that instant I remembered he'd been dead these twenty years, and I woke up. I woke up and rubbed my eyes and tried to blink away the sadness of never having made the grade in Dad's opinion, and how he'd never had the heart to tell me so, except with those sad looks.

And I'd never redeem myself to him, or be a man with a woman, or see Mara again.

Never, never, never—the word sounded like the lapping of muddy waves.

## River

*Hugh Dunkerley*

The whole thing is always slipping away
   downstream,
its sliding surface a welter of accelerations
   and sudden brakings,
of whirlpooling gullets.

Where the water rides roughshod over stones,
   sunken trees,
it roughens in foaming reversals,
flows upstream like another river,
   tussling with the current.

Whatever debris is bodied downstream,
   unweighted by the water's uplift,
is laved in the seamless swimming,
is turned and spun by watery hands.

Water boils at the legs of bridges,
is torn and maddened by boulders,
slides over weirs and shatters,
   thundering its applause.

# V

## *Rising Currents*

Leonardo da Vinci, *Implements Rained Down on the Earth from the Clouds,* ca. 1498.

# The Lost Notebook of Aqueous Perspective

*D. L. Pughe*

> *Describe all the forms taken by water from its greatest to its smallest wave, and their causes.*
> —Leonardo da Vinci, *Notebooks*

It became a deluge for Leonardo to write of water. All its forms were to be explored, beginning in nooks and crannies of ice, where it melted and pooled, flowing into brooks, streams braiding into rivers, rivers splashing into seas, and finally oceans rising in enormous crescendos from gargantuan storms. The studies of currents alone took up page after page, the exploration of clouds into rain a complete chapter, and the deep exploration of underwater vision: a lost notebook.[1]

He is in Cloux near the castle of Amboise in the Loire Valley. He had retired to the royal manor house at sixty-two after exhausting nearly every possibility in Florence, Venice, Milan. Here he is allowed to think and dream, and occasionally to help Francis I, the French king, with plans for a network of canals for the Loire.

Awake in the night, he goes over again and again in his imagination the outlines of forms he has been studying, the "noteworthy things conceived by subtle speculation."[2]

He considers the great amount of hidden water lingering inside the earth which feeds the springs. The motionless high mountain lakes and ponds, the stillness of fountains and stagnant pools. He attributes their calm to their distance from the center of the earth, and yet it is from these heights that splashing rapids, thundering waterfalls, and great rivers fall like drapery: the Ticino from Laggo Maggiore, the Adda from Lake Como, the Rhine emerging from Lake Constance and Lake Chur.[3]

Fire flutters upward; water trickles down or drops from the sky. He is fascinated by liquid gravity, always descending, falling over, plunging, flowing downstream, pouring into the sea. Fire draws what is caught in it toward the sky like a passion that consumes you as it lifts you; water pulls you under in its powerful embrace.

Was it this attraction as well as this fear that led Leonardo to contemplate it again and again, from every perspective?

He drew rivers the same way he drew veins in the arm, a confluence of streams joining together and coursing to the ocean, nourishing continents along the way. Reaching the sea, they unravel in countless capillaries, spilling their pure contents into salty waves. He is fascinated by these mergers and divergences, the sweet purity of springs and heavy salinity of seas. He studies the way water courses around objects in its path, experimenting with sticks and boards, drawing the curls and braided currents as it rushes past every obstacle. He sketches a smaller river bending into a larger one from the opposite direction, whirling as the currents adjust themselves to the same destination.[4] He realizes the ways water must be coaxed into channels and invited into canals by a beckoning drop in height.

Leonardo is aware that water is not gentle in return. In the long nights at Cloux, he dreams again and again of the massive deluge he is sure will wash civilization away. He sees clouds rise in ferocious gray armor and drop ton after ton of water on a helpless countryside full of desperate humanity. Mountains crumble, whirling waves rise, fly up, recoil, "friction grinds the falling water into minute particles, quickly converted to a dense mist, mingling with the gale in the manner of curling smoke and wreathing clouds," then all is washed away in a deadly inundation of foam, furiously rushing to the depths of the sea.[5]

He imagines the deep, hidden channels of water within the earth that, over centuries, have carved away hollow caves. He is torn by contradictory emotions of fear and desire: "fear of the threatening dark cavern, desire to see whether there are any marvelous things within."[6] Whirlpools sometimes stir there, awaiting each luckless thing pulled into their path. They can appear dark blue and calm, for instance, running among jagged cliffs behind an innocent Madonna cradling her child in a nest of rocks.

The exchange between oceans and rivers fascinated him, water constantly circulating and returning, the infinite number of times all the waters of the sea and rivers have passed through the mouth of the Nile.[7] "That which crowns our wonder in contemplating [water] is that it rises from the utmost depths of the sea to the highest tops of the mountains, and flowing from the opened veins returns to the low seas; then once more, and with extreme swiftness, it mounts again and returns by the same descent."[8]

Gradually these torrents of water "throw back stones toward the mountains," hitting one another and rounding their edges away. By the time they reach the sea, they are pebbles worn to sand. Just as the water itself is caught in eternal return, so the land goes back and forth over centuries:[9]

| | |
|---|---|
| *Li moti son fatti dalli cor si de'fiumi;* | Mountains are made by the currents of rivers, |
| *Li moti son disfatti dalli cor si de'fiumi.* | Mountains are destroyed by the currents of rivers.[10] |

Was it this that drew Leonardo on?

What leads him to draw a submarine, to take Aristotle's notion of a "diving bell" and want to plunge into the deep-violet depths of the ocean? Knowing his accumulated fears of water, of the deluge that could sweep humankind away in a thunderclap and swirling curl of foam, what did he hope to find in the fluttering light on the ocean floor?

On terra firma Leonardo was trained in the strict linear perspective of Alberti, a world controlled by point and line falling onto planes. Forms recede into the distance in graduated angles, always toward an anchoring point on a horizon, which is also the point of vanishing. The Euclidean geometry that lay behind Alberti's scheme was widely believed then and rarely challenged. In the beginning Leonardo applied angles to light and air, and saw the boundaries of shadows as determinable points. "He who is ignorant of these [points]," he warned, "will produce work without relief, and the relief is the summit and soul of painting."[11]

But in the shadowy, spiraling realm below the surface of water there is no such relief. Plunged below an anchoring horizon, all points are indeterminable. It is a realm of impossible angles where light dances like snakes rising from the floor instead of crisp beams slanting down from the sun. Perhaps he was headed there all along, toward a more inexpressible abstraction. In his late writings Leonardo begins to redefine a point as an instant in time and a line as a length or duration.[12] Gradually, the lines of a winding river come to resemble his own time on earth, ending in the immensity of the sea. Duration there becomes blurred, hinting at eternity.

On land, Leonardo devised a new means of atmospheric and aerial perspective to eloquently explore this blurring of the world. He noted the way objects diminish in sight as they recede from the eye, the way colors change in the distance (often merging into blue), and the haziness of edges, of vanishing—"the way that objects ought to be less carefully finished as they are farther away."[13] He applied these luminous rules to his painted landscapes, overriding hard-edged geometric schemes. And he delighted in the waters that trailed off into the bluish distance, in particular those at the edge of a certain portrait of a lady with whom he shared his room in Cloux. In the evenings did he linger in the deep turquoise water that travels through the craggy mountains behind her indecipherable smile?

He began to call the Albertian method "simple," a *construzione legittima,* but nonetheless artificial, then started to decode what he called "natural" perspective. He resurrected the importance of our rounded world. The alleged "pyramid of vision" that is said to frame nature and reach into our eye would give us a much different view from the one we actually see. How did this play out under the surface of the ocean? How did this man, for whom the edges of light were once calculable angles, manage in the restless coils of aquatic shadows?

He began to embrace curves. He observed that stones flung into the water become the center and cause of many circles, that sound also diffuses itself in the air in echoing rings. And he noticed the sky spreading out in concentric bands of atmosphere, with the horizon suddenly visible as a graceful arc.[14]

Peering into water from above, he finds that direct light does not allow him to look deeply into the layers of a stream, it bounces back his reflection and the sky. His eyes find a way in through other dark shadowy images reflected on the surface or what he calls the "skin." Observing submerged pebbles, he sees how light bends as it enters the liquid world then diffuses in unpredictable ways. It would be several hundred years before a "wave" theory of light came to be accepted, but he was already suspicious of the sanctity of beams.

Straight things seem to dissolve into ribbons in watery realms. Descartes would later describe how a stick in water appears bent from refraction, and that only a child would most likely believe it is truly bent. "Touching it, however, confirms that it is straight and goes beyond our preconceived opinions."[15] Descartes believed all "visual errors" could be corrected this way—our senses working furiously in tandem to check and balance one another. He calls this ability "reason"; Leonardo called it common sense, *senso commune,* where the five senses minister to the soul and enhance artistic perception.

He was the master of depicting shadows, smoky contours. *Sfumato* was his special means of rounding edges by subtle gradations to capture three-dimensional views. He explains how the density of a shadow is darkest closest to that which casts it, then fades away as it stretches into the distance. But looking down on water, shadows from above leap great distances and nearly disengage from their source. Bridges, for instance, leave their wobbly geometry on the surface some meters up or downstream. And immersed underneath the sea, the shadow of a fish swimming above can dart across the ocean floor a safe distance away just as its fin touches your shoulder. How did he try to capture these shifts of location, the curves of light, the diabolical dimensions of vision in the hushed immense chamber of the ocean?

We cannot know. In his rooms at Cloux in those final days he dreamed again of the deluge, his sketches becoming more and more cataclysmic and yet full of a vibrant conclusion: visual clashes of cymbals and drums and trumpets. He finds a sketch of a storm of our human failings, our foolish desire for objects suddenly bursting and falling from a dark thicket of clouds.

Comforting him in his last hours, the outstretched arms of St. Anne, the enigmatic face of La Gioconda, and the even more mysterious smile of St. John, his flamelike finger pointing toward the sky. They all appear to know something. And joining him in thought: a self-portrait, his beard cascading in silent waterfalls.

He had once believed that swimming is the closest we can come to understanding what birds do in the air. It can free us from gravity and fear. Like flying, it is a pushing away, and he believed we could push, that water could be mastered. Challenging Christ, he devised shoes with helpful poles for walking on the surface. And attempting to defy the gods, he designed a lifesaver not unlike those carried on giant ships traversing the seas.

But as his apocalyptic dreams grew, these inventions paled. He again and again is caught in the storm:

> The swollen waters of the river, already having burst its banks, will rush on in monstrous waves; and the greatest will strike upon and destroy the walls of the cities and farmhouses in the valley. . . . The swollen waters will sweep round the pool which contains them, striking in eddying whirlpools against the different obstacles, and leaping into the air in muddy foam; then, falling back, the beaten water will again be dashed into the air.[16]

He had once designed an underwater suit to dive the depths of the ocean. Did it occur to him now that the ocean floor might be the safest place to hide? He had altered his judgment of the world and had begun to embrace the curved reality we too have recently begun to know. What did the deepest depths of blue water hold in final reckoning? A resting place, a refuge, a realm of constant revision and invention. A point of vanishing where the soul is firmly anchored in scattered turquoise light.

## Notes

1. The notebook of underwater perspective was not mislaid, it was simply but sadly never written. It is not known if Leonardo did intend to write one.

2. Leonardo da Vinci, *The Notebooks of Leonardo da Vinci* (Oxford: Oxford University Press, 1952), p. 218.

3. Leonardo da Vinci, *The Notebooks of Leonardo da Vinci,* vol. 2, comp. and ed. Jean Paul Richter (New York: Dover, 1970), sec. 933, p. 181.

4. Ibid., sec. 971, p. 201.

5. Ibid., vol. 1, sec. 606–611, pp. 305–314.

6. Ibid., vol. 2, sec. 1339, p. 395.

7. Ibid., sec. 945, p. 187.

8. Ibid., sec. 965, p. 197.

9. Ibid., sec. 919, p. 175.

10. Ibid., sec. 979, p. 205.

11. Leonardo da Vinci, *Trattato della Pittura di Leonardo da Vinci,* sec. 121, quoted in Kenneth Clark, *Leonardo da Vinci* (Cambridge: Cambridge University Press, 1939), p. 76.

12. Leonardo da Vinci, Codex Arundel, sec. 190 in *The Literary Works of Leonardo da Vinci,* ed. J. P. Richter (London: Oxford, 1939; rev. ed. New York, 1970), para. 916.

13. Leonardo da Vinci, *The Notebooks of Leonardo da Vinci,* ed. Edward MacCurdy (New York: Reynal, 1939), p. 864.

14. Leonardo da Vinci, *Notebook A* (Paris: Bibliothèque Nationale 2038) R paragraph 69, quoted in James S. Ackerman's *Distance Points: Essays in Theory and Renaissance Art and Architecture* (Cambridge: MIT Press, 1991), p. 116.

15. René Descartes, *Descartes, Selected Philosophical Writings,* trans. John Cottingham, Robert Stoothoff, and Dugald Murdoch (Cambridge: Cambridge University Press, 1988), sec. 439, p. 124.

16. Martin Kemp, ed., *Leonardo on Painting* (New Haven: Yale University Press, 1989), pp. 234–235.

## Thoughts Breathing in a Blizzard

*Antler*

Breathing air with snow falling through it
   thinking how flour is sifted through a sieve
Each different snowflake design is a sieve
   and the air is charged with the energy
      of each snowflake design
         falling through it,
Air passing through the shapes
   of openings in the interior
      of each falling flake
   as well as along the edges,
Serrating in minute invisible architectures
   zillion-shaped clarified vibrancy
      of snowflake sculpture reality
   the edges of snowflake symmetry
      chiseling microscopically
   into an invisible display
      of snowflake-sculptured air
   beyond human comprehension.
And as an animal leaves its track
   in wet sand or mud
      so each snowflake design
   leaves its track
      in cool moist air
   as it falls,
So that in breathing air
   snowflakes have passed through
      you breathe the invisible tracks
   their designs have imprinted
      in the air
And as a hunter follows the bear
   by stepping in its tracks
      you follow the snow
   by breathing the invisible
      patterns of its designs
         pressed into the air
      as it falls.

Sally Gall, *Amazon,* 1986. Courtesy of Julie Saul Gallery

# The Music of Living Landscapes

*Wilson Harris*

Perhaps I should tell you that I studied land surveying and astronomy as a young man. That was really the launching pad for expeditions into the deep, forested rainforests of Guyana, so that I became intimately and profoundly involved with the landscapes, and riverscapes, of Guyana.

As a surveyor one is involved in mathematical disciplines, and astronomy, and one has, or I have, the sensation that the part of the cosmos in which we live, and the rainforests, are the lungs of the globe. The lungs of the globe breathe on the stars.

It seems to me that, for a long time, landscapes and riverscapes have been perceived as passive, as furniture, as areas to be manipulated; whereas, I sensed, over the years, as a surveyor, that the landscape possessed resonance. The landscape possessed a life, because, the landscape, for me, is like an open book, and the alphabet with which one worked was all around me. But it takes some time to really grasp what this alphabet is, and what the book of the living landscape is.

Let me begin my address by asking—

Is there a language akin to music threaded into space and time which is prior to human discourse?

Such a question is implicitly imprinted in legends of Guyanese and South American landscapes, preternatural voices in rivers, rapids, giant waterfalls, rock, tree.

A fish leaps close to where I stand on a riverbank, in the great dark of the South American rainforest night, and look up at the stars.

Theatre of memory! I hear that leap or voice of rippling water all over again across the years as if it's happening *now,* this very moment, within the Thames of London beside which I have often strolled since arriving in England.

And one meditates here by the Thames—as one meditated there not far from the Amazon or the Orinoco—upon the fate of the earth and its species.

Outer space is steeped in dangers and in environments hostile to life. Yet, it is said, the human animal is a child of the stars.

That child must carry, surely, an instinct for creation. Not only an instinct for the fury of creation, the fiery birth of constellations, but an instinct for immensity and silence, the music of silence within contrasting tone and light and shadow as they combine to ignite in oneself a reverie of pulse and heart and mind.

Inner ear and inner eye are linked to eloquent silences in the leap or pulse of light in shadow, shadow light, as if the fish in remembered rivers fly through an ocean of space and witness by enchantment, it seems, to the *miracle* of living skyscapes, oceanscapes, riverscapes wherever these happen to be, on Earth, or at the edge of distant galaxies.

I was born in South America and I left British Guyana for the United Kingdom at the age of thirty-eight.

Guyana is a remarkable wilderness. It has known Spanish settlers, then French and Dutch rule but became a British colony in the early nineteenth century.

Its population is less than a million but encompasses peoples from every corner of the globe, Africa, India, China, Portugal.

In area it is virtually as large as the United Kingdom and one sees graphically I think, on a map the two oceans, so to speak, that flank the narrow strip of coastland along which the greater body of the population live and sound their drums of India and Africa. One flanking ocean—with its subdued, perennial roar against sea-wall and sea-defences—is the Atlantic, the other is green and tall, unlit by the surf of electricity on rainforested wave upon wave of wind-blown savannahs running into Brazil and Venezuela.

There are Amerindian legends which tell of sleeping yet, on occasion, singing rocks that witness to the traffic of history, the traffiic of expeditions in search of El Dorado that Sir Walter Raleigh would have contemplated when he voyaged up the Orinoco before he lost his head in the Tower of London. The rocks sing an unwritten opera of El Doradonne adventurers.

Amerindians such as the Macusis, the Wapishanas, the Arawaks—whom Raleigh would have encountered—are still to be found in Guyana and South America. They have suffered—since Columbus's and Raleigh's day—decimations and the continuous depletion of their numbers at the hand of Europe across the centuries.

They were close to extinction at the beginning of this century but have survived in small numbers against all the odds.

I came upon them frequently—indeed they were sometimes members of my crew in landsurveying expeditions into the heartland of Guyana and was drawn to their still demeanour and mobile poise as huntsmen and fishermen.

Through them I learnt of the parable of the music of the fish in a rippling stream. They baited their fisherman's hook with a rainbow feather from a macaw or a parrot and with a twist of the wrist—as if they addressed an invisible orchestra—made it dart in the stream towards the leaping fish.

Feather from a wing and eager fish were united, it seemed, into an orchestra of species and a sacrament of subsistence they (these ancient peoples) had long cultivated since their ancestors emigrated twelve thousand years ago from Asia across the Bering Straits into the continent we now call America.

Emigration—in distant ages as in modern times—is the nerve of spiritual enterprise in all communities; it is driven by private necessity as well as economic and historical impulse, by hope, desire, promise and innermost vocation.

I emigrated to the United Kingdom in the late 1950s and lived with my wife Margaret in Addison Road, close to Holland Park, a stone's throw from Kensington Gardens.

It wasn't long before I resumed the orchestration of elements in fiction—which I had begun in the landscapes of South America—with the novel *Palace of the Peacock*. This was the first volume in a related series to be entitled *The Guyana Quartet* upon which I worked in the early 1960s.

I had few friends in England, none of influence, and at such times when the future is grave and uncertain and one is a stranger in a great city one is visited by archetypal and troubling dreams.

Archetypal dreams employ symbols of brokenness to depict the shedding of habit. A naked jar sings in a hollow body, sings to be restored, refilled with the blood of the imagination. The jar sleeps yet sings.

The jar is adorned with many elusive faces. It is inscribed with the head and the body of a boatman in a South American river. The boatman contends with mysterious currents. The paddle with which he strokes the rapids is seized by a streaming hand arising from the bowels of the earth . . .

An orchestra reawakens in my mind instinctive with a surge of terrifying music in the voyaging boatmen in *Palace of the Peacock* and I turn a page in the book and write—

> The boat shuddered in an anxious grip and in a living streaming hand that issued from the bowels of earth. We stood on the threshold of a precarious standstill. The outboard engine and the propeller still revolved and flashed with mental *silent* horror now that its roar had been drowned in other wilder unnatural voices whose violent din rose from beneath our feet in the waters.[1]

The *silenced* roar of the engine filled the boatman with dread. The engine was strangely alive in the void of his senses, void because in the heart of danger he could *see* its activity but no longer *hear* its voice as it assisted him and his crew as they paddled against the torrent . . .

But surely there is more to silence than a void in civilization? Is there not an inner music to silence, an inner attunement of ear and eye to sounding waters and painted skies, painted earth?

I took a night job for a few months in the winter/spring of 1960 in a North London factory.

On leaving the workplace early mornings I took away the hoarse call of the wheels of industry and the clamour and the grind of sliced metal as dawn broke like a white-feathered bird in the early spring blossom of a horse-chestnut tree at the gate through which the factory workers streamed to kick-start a motor bike or take a bus or drive a car. Layer upon layer of noise drowning noise.

Even as I moved through the gate in the stream of workers I knew the rain-forests in Brazil and Guyana were under threat, erosion of soils was occurring in the United States and around the globe, weather patterns were changing . . .

Nature is not passive. Nature erupts into orchestras of Nemesis.

Yet it *knows* our peril for we are in nature, of nature's chorus in response to hurricane or waterfall. Nature arouses us to speculate on orchestrations of inner eye and inner ear beyond every void of the senses, beyond every grave of the senses . . .

I left the factory in the Spring of 1960 imbued with a sensation of profound necessity in the life of the imagination to visualize links between technology and living landscapes in continuously new ways that took nothing for granted in an increasingly violent and materialistic world.

The haunting and necessary proportions of a new dialogue with reality in all its guises of recovered and revisionary tradition drew me into an anatomy or shared body everywhere in all things and species that give colour and numinosity to space. . . . Lazarus emerges in the first volume of a *Carnival Trilogy* which I wrote in the 1980s.

Inner ear and inner eye are his resurrected anatomy attuned to the music of painted silence in pulse and heart and mind arisen from the grave of the world.

He travels one winter day from his workplace on a bus that stops to pick up passengers close to a fence at Kensington Gardens.

He sees a carpet of burnt autumn leaves across the fence. A veil descends from a cloud in the sky shaped like a pinnacle of flood. The veil breaks into smoke and mist and runs into a hollow depression within the painted silence of the leaves that are eloquent with the immensity of changing seasons, contrasting tone and texture. Inner ear and inner eye weep for all things like the dew of rain. Lazarus is aroused to the beat of cosmic love as he contemplates the trampled leaves in the passage from *Carnival* I shall now read:

> In the winter light that seemed to echo with intimate yet faraway vistas . . . Lazarus felt the imprint of black fire, black tone, numinous wonderful shadow. . . . Yes, mind, heart, shadow . . . was the mind of Lazarus in attunement to ivories of sensation, russets, and other alphabets of the elements within every hollow epitaph of memory, every hollow grave.
>
> Winter lapsed into the carpet of autumn leaves under the bole of a tree that the bus was passing. The trampled leaves appeared to smoke with an arousal of spirit, trampled greenness, trampled yellow paint, in the hollow depression of time and place from which one arises to discourse with silent music within the roar of a great city.[2]

[Pause. Music that dies yet seems to echo endlessly in risen consciousness as in the last sound in Gustav Mahler's Ninth Symphony.]

Lazarus may have arisen in the early 1980s in *Carnival* but alas the grave of history continues to yawn wide even as I speak now in the year of Our Lord 1996.

The savage exploitation of rainforests continues in South America. Trees are felled like dumb creatures. River catchments are impoverished. The muse of nature within the consciousness of peoples is threatened. A deadly cyanide overspill seeped into the great Essequebo River of Guyana in 1995. The grave is deep despite every carpet of leaves that Lazarus paints with music.

When I speak of *silent music*—in the short passage that I read from *Carnival*—I am intent on repudiating a dumbness or passivity with which we subconsciously or unconsciously robe the living world. Living landscapes have their own pulse and arterial topography and sinew which differ from ours but are as real—however far-flung in variable form and content—as the human animal's. I am intent on implying that the vibrancy or pathos in the veined tapestry of a broken leaf addresses arisen consciousness through linked eye and ear in a shared anatomy that has its roots in all creatures and in everything.

Consciousness of self in others, consciousness of diversity that breaks the mould of prejudice, remains a mystery to science.

In mystery lie orchestrations of comedy that fuel the imagination to release itself from one-sided dogma whether in science or in religion. Astronomers have sought to fathom the age of the universe only to stumble in a comedy of space in which subordinate parts within the universe seem older than the parent universe itself. One awakens at times to one's frailty in the cradle of the mind in particles that settle on one's brow or hand or skin, sailing particles from distant mountains and valleys that seek their mysterious parentage in all substance or in the alchemy of sound in a rainbow.

Such, I believe, is the implicit orchestra, of living landscapes when consciousness sings through variegated fabrics and alternations of mood, consonance as well as dissonance, unfathomable age and youth, unfathomable kinships.

In an age of crisis the marriage of consonance and dissonance—transmuted into unpredictable and original art that challenges the hubris of one-sided tradition—is an important factor, I think, in the re-sensitizing of technology to the life of the planet.

The taming of the wilderness has always been an uneasy programme, however desperately pursued, in the body of civilizations.

When I come upon a felled tree in a park in England it sometimes shapes itself in my inner eye as the epitaph of a murdered forest in Brazil, or Guyana, or Venezuela.

I seek—as if imbued by Lazarus's mind in my mind, Lazarus's dream of cosmic love—to re-clothe that tree with the music of consciousness, with rustling, whispering branches in the foliage it has lost.

I picture the tools that felled the tree as newsprung branches themselves within a parable of creation which gives breath to iron or wood or rock. Adam was moulded, it is said, from clay. Thus the technology that killed the tree arrives or returns as living branches in the risen tree itself.

Each newsprung branch—whether wood, or iron or stone—sees itself now as susceptible to a more deadly invention or tool than it had been when it felled the kinship, resurrected tree, parent or child, to which it has returned. The risen tree in my consciousness, veils flesh-and-blood into itself within a revisionary dynamic of creation and re-creation. Even as the technology of clay was moulded in genesis into Adam's pulse, Adam's breath.

Cities have come to nestle in branches of clay or stone in valleys or mountains. They too may be revisited with an inner eye to see how vulnerable they are. Their hope is born of the life of imagination's tree in which sculptor and painter and architect and carpenter and mystic sensitize and re-sensitize themselves to rhythms and pulses orchestrated through being and apparent non-being.

In such a re-visionary muse, or music of consciousness, the tree suspends itself, promotes itself by degrees, within theatres of crisis that might be seen or read or gleaned through a variety of perspectives . . .

The sleeping yet singing rocks of ancient Amerindian legend grow in that tree . . .

I hear them as I stand in the open parkland in Kensington Gardens half-a-mile or more from the orchestra of the traffic arriving at, or coming from, Marble Arch . . .

Where is Marble Arch? In this instant I am uncertain for I am drawn back in memory across decades to the Tumatumari falls in the Potaro River of Guyana not far from the place called Omai where the cyanide spill from industrial workings entered the great Essequebo River in 1995.

*Tumatumari* is an Anglicized word drawn from an Amerindian root text and it has been translated in different ways to mean "womb of song" or "sleep of song" or "sleeping yet singing rocks" in a Dream time, Dream space.

I visited Tumatumari in the mid-1940s to investigate hydro-electric potential in the vicinity of the falls.

I lay in my hammock at nights listening to the falls: hoofs of horses were running in the night as the rapids drove upon, slid over, pounded on rock: the high heels of invisible women clicked on a pavement and approached from the chasm of the river: there was the sudden grind of gears, there was motorized traffic: there was the back-firing of a car at Marble Arch . . .

The wind changes over Tumatumari as it changes over Bayswater Road and in the orchestra of memory and place which tunes into distant rain in a forest. So far away it sounds like the stealthy, drizzling approach of fire or a flute-playing ghost in the elements, a violin fashioned from sudden, scarcely visible lightning or metallic storm.

The rocks perform a tidal function in the scaffolding of the traffic of rivers. A fantastic balance between conservation and discharge of resources is achieved in the sculpture and placement of the rocks.

In my years of surveying rivers I was drawn into the sensation that sleeping and singing rocks are also dancers (stationary as they seem) even as trees and plants are known to walk under the close scrutiny of science.

The phenomenon of apparently immobile rocks which play a tidal role in non-tidal rivers is a miracle of evolution. Non-tidal rivers run ceaselessly downwards from their headwaters or sources in a distant watershed. They lie above the reach of the ocean tides which cannot therefore exercise a check upon their volumetric decline and discharge. They would become a huge empty ditch or trench were it not for the sculpture and placement of living rocks—their shape, wave-sculpture, escalation, placement in the fury of the rapids—within great river systems. They dance an inner, staggered, relay dance subsisting on the volumetric ball of the river that they bounce from hand to foot in their guardianship of resources, in their cultivation of the mystery of freedom and passage through diverse channel. They run and dance without appearing to do so even as the tree of life carpets the ground yet rises and walks in the limbs of Lazarus.

The body of the dancer in a living landscape is the technology of music.

The body of labour in living vocation resides in the technology of the resurrection.

Consciousness attunes itself to living landscapes within the dance of a seed in the soil. . . . From such a seed great cities grow and their echoing tracery is in the fall of a feather from the wing of a bird.

In that feather is the technology of space fused with the murmur of threatened species that still arise and address us . . .

## Notes

*Unabridged version of typescript submitted to BBC Radio 4. An edited version was broadcast November 12, 1996. Sound effects, flute variations were by Keith Waithe.*

1. Wilson Harris, *Palace of the Peacock* (London: Faber and Faber, 1960), p. 21.

2. Wilson Harris, *The Carnival Trilogy* (London: Faber and Faber, 1993), pp. 140–141.

*Bluefields, Nicaragua.* © Helen M. Ellis

# Aquifer

*Pamela Ryder*

Some summers just a stick of buckeye starts you walking.

Some nights in spring a moonrise and the tug of far-off tide will be enough to steer a peeled-down switch of birch or start a wobble at the wrist. Enough to pull a branch of tupelo clean-stripped of petiole and leaf, held bud tip down to lead you to the seep beneath the Little Dipper. Kept thumb up at the fork to feel the pulse of sap, the sense of going somewhere—of turning of a wing of mountain ash above the water, or sinking where the pickerels face side by side into the flow and scarcely wave a fin or raise the silt. Of following behind the shadbush bloomed too early for the upstream swim of bluegill, or carrying the sprig of full-moon maple to the place the aquifer has had its fill. Seeking the spot a brook that likes to hide itself will reappear roadside, ditchwise, lit with foxfire bright as someone's lantern or camp. Finding the pool above the ford where we pitch the pebbles in: where we stir the whale-spout spray of constellations and scatter the lights outside of Orion, the stars poured from the jug beside the Water Bearer's feet.

Inside something sprouting pale and spindly pushes past the cistern, climbs up the cellar stairs, and sends a tendril to the shelf where we put rows of jars of harvest years put by. The hard-pressed garden sends up only straw-grass and crab; grows only dust devils, dogbane, poorman's pepper. But still the Allegheny plum spreads a branch above the porch to border pieces of the blue-boughed window view of sea, and there it shades our days.

We count the days. We count the clouds blown by and the watch ticks it takes to hear a stone sent down the well meet water. Days we watch the wisps of cirrus, mare's tails, but never the cumulonimbus that shifts into an anvil shape, a pile-up dark and detonated. No peals are heard within the heap that rolls in thickly undersided, bolted, spooking the foal into an unbroke bucking up and down the line of fence. No foul-weather forecast or signs of precipitation. No predictions of a squall blown in from the sea. No drop in the belly of the glass-blown barometer that lifts the meniscus up the spout and sends the rock doves sheltering in eaves and evergreens, then starts the dogs snapping the strands of grass that the sickle missed around the borders of our yard.

Just skies of mackerel and buttermilk. Just dog days the noon comes calm and clabbered, birds mute, bivouacked in shady bottomland while we spend our nights beneath the Little Bear, the Dog Star.

Sunday mornings with no wind prevailing, the dogs are porched early or under, and we hear the short-cut sermon. We bow our heads, hoping for a downpour and hearing that the pharaoh's sky is overflown with locusts. We make our prayers for weather while Noah is awash somewhere and the Nile is turned to blood—dried up or undrinkable. All please rise, we hear, and comes the closing hymn: husbands with hat in hand, straw and weave; and wives in bonnets for the sun or ones close-brimmed in roses of unwilting silk. While all the while they fan their fussy babies—spitting, gassy, bibbed, and tuckered—little ones and converts gussied up and overheated for the most holy dunk in the spirit, for their baptismal dip.

Sunday after suppers, fresh-washed cups and plates are stacked upon the shelves, cupboards are shut, crumbs swept up. The men sit shirt-tailed out, talking of crops, cows, yearlings; speaking of the turning of the year; speaking of the leaves turned early. Turning the horseshoe on the springhouse door end up in the hope that somewhere a front is holding back, or tucked in and holding tight. Hoping that the stick of Chinquapin they always gave the child they picked to hold for luck would whisper to her; would tell her what the wind might say before a rain; would pull her, point.

A catkin of mud-banked cottonwood was the child I was. "Sopping wet!" the women said who picked me up, always puddled. No matter where they set me down—on quilt or grass or hard-packed yard—crawdad and peeper always found me when they found me with my fisted sprig of chestnut or my stick of cherry wood. Dripping wet they took me from the spring that no one knew was there before—but must have been, while everyone remembers how it bubbled up! The font that everyone remembers overflowing when it came my time for naming, my turn for dipping in the name of, while all the other babies stayed sleeping or sweetly screaming. Me—kicking off the swaddle from my feet, splashing hand and heel against the bowl of stone. Spilling the sacramental waters of immersion. Flinging drops into the light, as blinding as the bolt-lit firmament above the road into Damascus, leaving wayfarers saved, onlookers soaked, and me newly sanctified.

The choir hushed. Rain slowed to a patter. The drainpipe singing. Pockets emptied. Silver dollars uncollected on the plate.

Rungs of sun. Roof slats bright and steaming. Plover chipping in the burr reed. Perch and rainbows rising near the high-water ford. Grass bent low under its starry weight.

These are days when cloud breaks make a Jacob's ladder: after a summer squall or red sky at morning, or just before a storm moves on above the sea, above the shore I scooped out in cups to let the water find me. Where I filled my pail with shells of moon snail and sand dollar, of cherrystone and lightning whelk and limpet. Where the surf-flung stars crawl back into the waves, and tiny coil-tailed horses are up-beached and brittle, sitting finely ribbed in the sand, reined seagrass tight, and winged where you might think would be a fin. Caught and corralled forever in a jar and carried home with other pieces of the sea, to be set on the cellar shelf beside the jars of butter beans and snap. Beside the Allegheny plum put up, the coltsfoot jam put by, the store of succotash, of chard, of stoneless cherries, and the horseshoe crab left pantry stranded and hoping for a tide.

High and low, and back I came, for a taste of the sea, for cockles and mermaids' purses. For beach plums collected in my pail and a look at the sea-logged sailors, muscled and singing a ditty. Down from the horse latitudes, up from the Dry Tortugas; out of the doldrums and Sargasso Sea: scrimshawed shipmen blown in with the seadoves, stepping lively to a seapipe whistle. Deckhands and seadogs set loose along the docks, unloading the oil of sperm and blue; taking on provisions and barrels of sweet water. The coxswain is coming with coconuts and tamarind. The bosun arrives with trinkets of ebony locked in his dove-tailed chest of teak. The high-water helmsman is taking his shore leave, testing his land legs, wanting me to be the storm he's riding out. And last, the foretopman, toting a sextant and taking liberties, talking of

Antares and azimuth, telling celestial longitude, sidereal time. Showing me ascension and declination. Demonstrating the perpendicular to the elliptic. Horizontal parallax. Planetary positions. Finding his bearings of bow-and-beam in my bed above the porch, in the room with the blue wide pane of plum-bordered window, above the clock-and-church-towered town, above the bright road crooking hill by hill on down to the river, above the fenced-lined fields sunburned up, or rain-flat down and a bough-framed view of the blue sky bay.

Who will bring me a bright stick of blossoms?

Who will climb from the upstairs window? Who will step out on the roof slats made of cedar, cut from the groves that grow along the river? Who will reach out to the bough that is slim enough for whittling?

Who will lean into the topsails of the tree, reach past the petals, and roust the doves from their nightfall perch? Who will bend the branch, cut the shreddy bark, twist it round and round the way a branch won't go and give me what I need to go where the water wants me to be going?

What shall I need to be going?

A moonrise and the right branch of tupelo held not too tight. A twig of Allegheny plum or crab-fruit apple. A wand of sandbar willow. Of horse chestnut. Of stone-fruit cherry.

Who will cut the whippy stick of hickory to pull me to the place a spring has been keeping secrets in seepy hillside?

Who will break the stem of buckeye that will take me where I know a well is waiting to be dug, where I can tell how deep the roots of a sea blue spruce or sycamore will reach?

Who will snap the bough of peach to say where a river leaves behind its mud shard bed and shows me the places it would rather sleep?

Not the midshipman who wants me battened down and bedded in my bed above the porch; not the one who brings me pomegranates wrapped in paper or the spiced plums in a jar. Not the able-bodied seaman who blows into port on a gale-high wind and spins back to sea on a waterspout. Not the one who brings for my supper a brandied pear still grown to its stem and slipped past the neck of a slim-necked bottle. Not the mariner with his pocket full of frippery and gimcrack. Not the middy swigging from the flask he filled with cups of sea at the crookedy Straits of Magellan. Not the mate who is looking for a good-time maid, or the roustabout wanting a sweet-water damsel. Not the cook who wants me keeping my cupboards and tide tables tidy swept, or the tar or gob or crew. Not the sea-faring fare-thee-wells watching out to sea from my window; waiting for a fair-sky day and a stiff breeze out; wishing for the empty spars to soon be sprouting sails.

Inside of me is something sprouting: tendril tailed and gilly.

Inside of me is the tickly swarm of silver fishes I see flying from the spindrift spray to see me.

Sweet flag and cocklebur are hiding bluegill and minnow.

Flags are run up for getting underway; sails are full and ready.

Downriver fingerlings are swimming past where reed and bulrush bend, slipping along the bow that takes away a landlocked sailor.

Inside of me is something sprouting: coil tailed, finned, and slit gilled.

Upriver in me something swims the neap and spring of tides, sways with the surge of salt. Inside something is timed to the bloom of the shadbush, to an upstream swim and spawn. It swims from me too early. It is born in my bed, in sight of the windowed sea; flowing plum tinted in the spill, in the flume of a tide turned red.

Pour it to the tide race. Spill it to the current of the ebb, before a passing shower starts the perch and rainbows rising, has them pulling at the unnamed pieces of me, mistaking parts of me for mayfly or caddis fly or freshwater damselfly or stonefly nymph. Spill it down from a jar on the riverbank. Pour it where the whirligigs will circle in the eddies and the riverbed bottom makes a bowl of stone. Turn the rocked-in pools blood-tinged and undrinkable. Empty the jar or flask or cup or colt skin bottle full of what you have been saving, or trying to save and sanctify me. Dry me up.

First crescent moonset will start the peepers singing.

Last quarter rise starts the dun foal in his sleep.

Some nights in summer there is salt-tanged wind and sea smoke moving on the water. There are sea-made clouds that hold in tight the heat of the day.

The church is dark and emptied out of singing. The doxology is done, the font dried up. Sleepy dogs stay guarding at the edges of the yard, yawing and yapping at who it is comes walking through the garden where the straw-grass bends at the sickle blade; at who it is comes walking through the full-moon maples. Take along a lantern. Hang it handle-notched to a sapling to see where you are headed, to see your way along the crooks in the road. Pick the tree that's strong enough to pull you, and straight enough and green enough to bend.

A whittled length of locust or Allegheny plum will have you stepping lively. A piece of catalpa or dogwood will certainly do. It will pull you to the place to dig if you should want to sink a well or build a springhouse. It will show you the spot to wait if you would rather have the water come to fill your pail or find you. It will take you where the cress has sprouted and the morning dewfall fills a seep. It will be the way that is lit by an early hour spray of stars set loose from a wave unraveled. A lodestar course leaning to Cassiopeia, bearing to Pegasus, steering through the constellations, flying past flying fish and keel and compass. Winging past seadove and dolphin from the depths of the heavens.

It will be a reading of meridian.

A starboard-side observation of degree and minute.

A plumb line to the vertical. A pitch and roll with the deck of a vessel, a moon-fixed correction for the aberrations of a dream.

It will be a dead reckoning past the stations of the sky that sleep the denizens of water.

A passage through the straits of the celestial sea.

## Jottings Inspired on Perusing Cantor's Transfinites

*Will Alexander*

*The essence of mathematics is its freedom.*
—Georg Cantor

*So far as the theories of mathematics are about reality, they are not certain; so far as they are certain, they are not about reality.*
—Albert Einstein

*. . . the freedom to explore what the mind wishes to explore . . .*
—Morris Kline

I think of lumen which leaps by elevated density
by a burning numerology of fractals
above a gulf of ordinal data risen to continuous intensity

a zone of bloodless transfinites
as is fire in riveting neon abstraction

as if
in dissertation on lagoons
water dispersed & increased by contraction
by mysterious conduction within a sullen flash of windows
or a compacted neutron counting
or glassy anodyne verbs
or a greenish horse with its mane in alacritous gallop
across a span of Uranian nightmare foundries

by desert
by essential extract replication
by light in its power through stamina

poised in its count by sunless anagrams
by attraction in carnivorous subset
condensed in the flameless motion of scarabs
like ascendant number in implosive turquoise sand much
like integral re-implosion
or a graft
or a "complex variable" as a fact of pure existence

I'm thinking of infinite parallel drafts
of beauty as fundamental derivative
as if a cylinder of mass were derived from a source
independent of aggregates
of analytic compound

the mathematic condensing of shadow
as subtracted primer
as sudden arbitrary spiral
as a mind spinning above a mecca of gates
inside the grasp of unplanned alembics

configurations
scattered across the trance of priority
being fire
being kinetic nettles & compounds
claimed by withdrawal
spun by a force of dazzling ascent

I think of the wells which are transfigured
by saltation
by proof as incalculable paradox

for instance
a maze discontinuous with origin
or alive with bursts of fulminate x-ray carrion
or sands divided & transfused by in-cellular distance

as if tracing a flame as seizure
then watching camels ascend through a stark Arabian sea
then witnessing metal as ray
as oblong with vivacity
like mirage at the brink of incessant mobility

& so equation more porous than salt
than various crystals & ices
become a fleeting source of solar bursts
by impeccable registration
by tectonic retro-cause

much in the way infinity erupts
by quasi proof as eclectic summons
by compression
by participatory drift
opened as dialectical saffron

therefore
Cantor as interior hurtling of sonar
of fulminate spells within balance
as if
on balletic equators
signals of stormy errata
of exploded aqueduct foundations

as if
equations could be conducted by spittle
by interior marionettes
by sistrums so singular that a rock from ovarian water
transmuted by pulse as culpable measure
as orational pleurisy
as power which implants frustration
so that
it subducts & supersedes
its modes
its chemistries
its fragments
its flailings at simultaneous boundary

## Spring Flood

*Dan Stryk*

Sometimes in the first melt, mixed with slow rains, it would flood. Then we'd have our terrible fun, chanting the name of our Indian river. Frenzied demons just released from the ice-locked zone of an Illinois winter, where it seemed that nothing flowed or moved but the distant voices of the crows, piercing the mists that covered them above our small town park.

There they'd hunch like aged parishioners whose shrill and useless bickering would vanish once again into gray mist, where nothing changed. And townsfolk seemed to feel that nothing should. Until the ice broke, groaning squeaky, on those first warm days. Then suddenly the park our homes surrounded was a vibrant, pulsing planet, thrilled birds circling above the swollen puddles turned to silver lakes in morning light of the first flood. Like a new earth or century, small rivers raging down our streets—the lunar cars, now strangely beautiful, nearly submerged.

Above the sodden basements, newscasts whined distraught through the thin walls of clapboard homes. Whined all day where townsfolk cringed like nervous wasps behind the doors they'd bolted for no reason but some ghostly fear, as basements rose each day to festering ponds.

But nearly naked, we'd run out on weightless feet—despite their warnings of the drowned—down rushing streets into the wondrous planet of the flooded park. Black willows probing, like mysterious signs, from primordial depths of the Kishwaukee's submerged banks. The crows, turned pterodactyls, cawing in black rings that circled, spying on us now.

We'd felt but vaguely troubled, then—vaguely that we'd spurred the change by some sly need, or unsaid curse. Yet hardly conscious of our ruined homes, the cost to come that worried all the world: those sober waters that it seemed the world could not control or understand. The dread, we were supposed to feel, that lay beneath the beauty of those waters we had no wish to control. Beneath the change we'd not feared, but embraced.

**Soaking Paper**

*Christopher Funkhouser*

Adam David Clayman

*make miles*
*of highway &*
*sickness of*
*civilization*
*"worth it"*
*by sun,*
*wind*
*hours by*
*water*
*removing*
*eye glasses,*
*worries,*
*impatience.*
*zen is this*
*everyday.*

Preparations: loading tools, carrying them,
conversation, soaking paper, taping frame,
mixing gum arabic with titanium dioxide.

to Paint

Dipping brush in sand—
glue to make it stick. Fingertip
application, getting sky right,
even infusion of water—

everything

beach, bay, gray skyline

cooperation, not control.

big boat.

bigger sea. little ones by the shore.

toward<br>
gentility<br>
below<br>
edge<br>
of clouds

Buzzard's Bay. On it, near it, in it<br>
always,<br>
glowing water—<br>
Sandy Rock

inlet's motion<br>
makes organisms sparkle<br>
bioluminescent. skyward<br>
stars flicker, shoot<br>
between fireflies<br>
time of year

oar, crab, fish, hand ignite<br>
jellyfish in palm such a sight (site?)<br>
to be held, harmless particles lustrous<br>
a million tiny lives alight

stars beneath water's surface<br>
a million tiny lives alight<br>
nature exploding in<br>
darkness of time as<br>
yesteryear

reflection lit by<br>
old house, voice<br>
can make it<br>
kinetic<br>
almost a reflection

"water light"
space below surface
mapped by molecules

how long can you look a

shot sky

long tail    catch

a flare

or fireworks

too high

to share with gnats

entering earth's sphere

spitting constellations

no moon

only to speak of

fish jumping food chain

underwater sparks

off of fingertips

Jaanika Peerna

# My Life with the Wave

*Octavio Paz*

*Translated by Eliot Weinberger*

When I left that sea, a wave moved ahead of the others. She was tall and light. In spite of the shouts of the others who grabbed her by her floating skirts, she clutched my arm and went leaping off with me. I didn't want to say anything to her, because it hurt me to shame her in front of her friends. Besides, the furious stares of the larger waves paralyzed me. When we got to town, I explained to her that it was impossible, that life in the city was not what she had been able to imagine with all the ingenuousness of a wave that had never left the sea. She watched me gravely: *No, her decision was made. She couldn't go back.* I tried sweetness, harshness, irony. She cried, screamed, hugged, threatened. I had to apologize.

The next day my troubles began. How could we get on the train without being seen by the conductor, the passengers, the police? It's true the rules say nothing in respect to the transport of waves on the railroad, but this very reserve was an indication of the severity with which our act would be judged. After much thought I arrived at the station an hour before departure, took my seat, and, when no one was looking, emptied the tank of the drinking fountain; then, carefully, I poured in my friend.

The first incident arose when the children of a couple nearby loudly declared their thirst. I blocked their way and promised them refreshments and lemonade. They were at the point of accepting when another thirsty passenger approached. I was about to invite her too, but the stare of her companion stopped me short. The lady took a paper cup, approached the tank, and turned the faucet. Her cup was barely half full when I leaped between the woman and my friend. She looked at me in astonishment. While I apologized, one of the children turned the faucet again. I closed it violently. The lady brought the cup to her lips:

"Agh, this water is salty."

The boy echoed her. Various passengers rose. The husband called the conductor:

"This man put salt in the water."

The conductor called the Inspector:

"So, you've placed substances in the water?"

The Inspector called the police:

"So, you've poisoned the water?"

The police in turn called the Captain:

"So, you're the poisoner?"

The Captain called three agents. The agents took me to an empty car, amidst the stares and whispers of the passengers. At the next station they took me off and pushed and dragged me to the jail. For days no one spoke to me, except during the long interrogations. No one believed me when I explained my story, not even the jailer, who shook his head, saying: "The case is grave, truly grave. You weren't trying to poison children?"

One day they brought me before the Magistrate. "Your case is difficult," he repeated. "I will assign you to the Penal Judge."

A year passed. Finally they tried me. As there were no victims, my sentence was light. After a short time, my day of freedom arrived.

The Warden called me in:

"Well, now you're free. You were lucky. Lucky there were no victims. But don't let it happen again, because the next time you'll really pay for it . . ."

And he stared at me with the same solemn stare with which everyone watched me.

That same afternoon I took the train and, after hours of uncomfortable traveling, arrived in Mexico City. I took a cab home. At the door of my apartment I heard laughter and singing. I felt a pain in my chest, like the smack of a wave of surprise when surprise smacks us in the chest: my friend was there, singing and laughing as always.

"How did you get back?"

"Easy: on the train. Someone, after making sure that I was only salt water, poured me into the engine. It was a rough trip: soon I was a white plume of vapor, then I fell in a fine rain on the machine. I thinned out a lot. I lost many drops."

Her presence changed my life. The house of dark corridors and dusty furniture was filled with air, with sun, with green and blue reflections, a numerous and happy populace of reverberations and echoes. How many waves is one wave, and how it can create a beach or rock or jetty out of a wall, a chest, a forehead that it crowns with foam! Even the abandoned corners, the abject corners of dust and debris were touched by her light hands. Everything began to laugh and everywhere white teeth shone. The sun entered the old rooms with pleasure and stayed for hours when it should have left the other houses, the district, the city, the country. And some nights, very late, the scandalized stars would watch it sneak out of my house.

Love was a game, a perpetual creation. Everything was beach, sand, a bed with sheets that were always fresh. If I embraced her, she would swell with pride, incredibly tall like the liquid stalk of a poplar, and soon that thinness would flower into a fountain of white feathers, into a plume of laughs that fell over my head and back and covered me with whiteness. Or she would stretch out in front of me, infinite as the horizon, until I too became horizon and silence. Full and sinuous, she would envelop me like music or some giant lips. Her presence was a going and coming of caresses, of murmurs, of kisses. Plunging into her waters, I would be drenched to the socks and then, in the wink of an eye, find myself high above, at a dizzying height, mysteriously suspended, to fall like a stone, and feel myself gently deposited on dry land, like a feather. Nothing is comparable to sleeping rocked in those waters, unless it is waking pounded by a thousand happy light lashes, by a thousand assaults that withdraw laughing.

But I never reached the center of her being. I never touched the nakedness of pain and of death. Perhaps it does not exist in waves, that secret place that renders a woman vulnerable and mortal, that electric button where everything interlocks, twitches, straightens out, and then swoons. Her sensibility, like that of women, spread in ripples, only they weren't concentric ripples, but rather excentric ones that spread further each time, until they touched other galaxies.

To love her was to extend to remote contacts, to vibrate with far-off stars we never suspect. But her center . . . no, she had no center, just an emptiness like a whirlwind that sucked me in and smothered me.

Stretched out side by side, we exchanged confidences, whispers, smiles. Curled up, she fell on my chest and unfolded there like a vegetation of murmurs. She sang in my ear, a little sea shell. She became humble and transparent, clutching my feet like a small animal, calm water. She was so clear I could read all of her thoughts. On certain nights her skin was covered with phosphorescence and to embrace her was to embrace a piece of night tattooed with fire. But she also became black and bitter. At unexpected hours she roared, moaned, twisted. Her groans woke the neighbors. Upon hearing her, the sea wind would scratch at the door of the house or rave in a loud voice on the roof. Cloudy days irritated her; she broke furniture, said foul words, covered me with insults and gray and greenish foam. She spat, cried, swore, prophesied. Subject to the moon, the stars, the influence of the light of other worlds, she changed her moods and appearance in a way that I thought fantastic, but was as fatal as the tide.

She began to complain of solitude. I filled the house with shells and conches, with small sailboats that in her days of fury she shipwrecked (along with the others, laden with images, that each night left my forehead and sunk in her ferocious or gentle whirlwinds). How many little treasures were lost in that time! But my boats and the silent song of the shells were not enough. I had to install a colony of fish in the house. It was not without jealousy that I watched them swimming in my friend, caressing her breasts, sleeping between her legs, adorning her hair with little flashes of color.

Among those fish there were a few particularly repulsive and ferocious ones, little tigers from the aquarium with large fixed eyes and jagged and bloodthirsty mouths. I don't know by what aberration my friend delighted in playing with them, shamelessly showing them a preference whose significance I prefer to ignore. She passed long hours confined with those horrible creatures. One day I couldn't stand it any more; I flung open the door and threw myself on them. Agile and ghostly, they slipped between my hands while she laughed and pounded me until I fell. I thought I was drowning, and when I was purple and at the point of death, she deposited me on the bank and began to kiss me, saying I don't know what things. I felt very weak, fatigued and humiliated. And at the same time her voluptuousness made me close my eyes because her voice was sweet and she spoke to me of the delicious death of the drowned. When I came to my senses, I began to fear and hate her.

I had neglected my affairs. Now I began to visit friends and renew old and dear relations. I met an old girlfriend. Making her swear to keep my secret, I told her of my life with the wave. Nothing moves women as much as the possibility of saving a man. My redeemer employed all of her arts, but what could a woman, master of a limited number of souls and bodies, do, faced with my friend who was always changing—and always identical to herself in her incessant metamorphoses.

Winter came. The sky turned gray. Fog fell on the city. A frozen drizzle rained. My friend screamed every night. During the day she isolated herself, quiet and sinister, stuttering a single syllable, like an old woman who mutters in a corner. She became cold; to sleep with her was to shiver all night and to feel, little by little, the blood, bones, and thoughts freeze. She turned deep, impenetrable, restless. I left frequently, and my absences were more prolonged each time. She, in her corner, endlessly howled. With teeth like steel and a corrosive tongue she gnawed the walls, crumbled them. She passed the nights in mourning, reproaching me. She had nightmares, deliriums of the sun, of burning beaches. She dreamt of the pole and of changing into a great block of ice, sailing beneath black skies on nights as long as months. She insulted me. She cursed and laughed, filled the house with guffaws and phantoms. She summoned blind, quick, and blunt monsters from the deep. Charged with electricity, she carbonized everything she touched. Full of acid, she dissolved whatever she brushed against. Her sweet arms became knotty cords that strangled me. And her body, greenish and elastic, was an implacable whip that lashed and lashed. I fled. The horrible fish laughed with their ferocious grins.

There in the mountains, among the tall pines and the precipices, I breathed the cold thin air like a thought of freedom. I returned at the end of a month. I had decided. It had been so cold that over the marble of the chimney, next to the extinct fire, I found a statue of ice. I was unmoved by her wearisome beauty. I put her in a big canvas sack and went out into the streets with the sleeper on my shoulders. In a restaurant in the outskirts I sold her to a waiter friend, who immediately began to chop her into little pieces, which he carefully deposited in the buckets where bottles are chilled.

Sally Gall, *Swimmers #2*, 1978. Courtesy of Julie Saul Gallery

## Water/Theos

*Norman Weinstein*

How it erodes the Stone Tablets of the Law, biblical, poly-,-economic. A Gnostic gospel, as yet unwritten, would claim Jesus thrives in the predictable display of baroque droplet dalliance. But all mysticisms, inevitably, invent their own Tracts of the Law, stone rough approximations of ThouShaltNots, against which pure spring water erases strictures. There is a watery freedom of an uncrucified savior, savor him, in the rising waves threatening, happily, to overwhelm pale illusions of harbor.

Lao-tzu watches his feet pool into a translucent pool where minnows swim. There's no terror. Eyes of his soles look centuries ahead to the filthy groundwater of New Jersey. He speaks a word that could imply "purity" but emerges instead as some scavenger's cry, lifting the minnow into a cloud-kingdom, perhaps a Middle Way, perhaps necessity.

A YMCA offers a "Terrified of Water Class," a swimming class for novices, appropriate for myself & anyone terrified of being swept away by inner/outer floods. What hilarity it must bring Poseidon, this collective therapy to render him postmodern, harmless as a trope. He's also selected the first initiate to drown in the pool's shallow end by forcing the beginner to hold his breath too long underwater for even a mockery of an epic.

This space is reserved for the kind of hydraulic engineer Debussy would have been—had he understood the nature of how water surfaces the denied feminine.

My oldcountry babushka grandmother stands on a Jersey pier, casts her rod into the gray Atlantic, reels in my entire oceanic vision, & drops it, a flapping-for-its-life flounder, into my unconscious hand. Close my fist around the vision, impulsively, city boy with no ocean sense, & drop it in my pant's pocket, removing it only when my MerchantMarine father ships out permanently. Particulars within that vision are long gone. Just a cadence re/sounding in a whiplash line on the page. From the inner ear of a shellacked shell comes her watery Yiddish, some sharp spiky surf resonance suggesting the futility of shipping out to escape the sinking dollar.

When the last kitchen tap sings taps . . . when the last live river goes home to its maker . . . still dream of rivers out of Eden invented to postpone the inevitable.

There's a sense where kidney patients prophecy the water's sickness, their cells cry out in despair as mercury roams the Pacific floor, silver tailings supernaturally glimmer in riverbeds. Because the pain of admitting this connection is too searing for the bodypolitic, every kidney donor becomes a water zealot by necessity. The angel knowing only the American English of "Piss on it" teaches this mantra to the not jaded.

A nineteenth-century Midrash tells that at the moment when God parted the waters of the Red Sea, every body of water on earth in that same instance divided. Who can argue the theology? The waters are still bifurcated—only we have no eyes for such nonsense now & live on & within the right side of the waters.

Idaho's Snake River owed nothing to snakes. A Bannock warrior made a snake-like "s" with his hands to warn an Eastern missionary of the river's hazardous turns. Misinterpreted as meaning "snake." Typical of such types to confound complex watery twists & turns, imposing their desert nomadic reptilian paranoia.

The child I've never fathered is tiptoeing into the Delaware's waters this morning. The child I've never been waves to him from the opposite shore. The Delaware can be named a "commercial artery." This is about the most authentic commerce I ship between the between of regrets.

Sounds of ice breaking on the Hudson River. Annea Lockwood composed the Hudson's ripping flowing sounds into a coherent musical composition—but the booming ice melts demand another kind of composer, musician, ruthless, dispassionate, an operatic tenor even, who, losing his voice to age & drink, for one last time remembers as he takes an inbreath how he nearly drowned in Venice as a consequence of an out-of-control prank on a crumbling bridge. His first note cracks like that Venice bridge held by deceptively gummy grout, false memory.

No river steps into the same person twice. As no river god knowingly does. In fact no river flows in any of these words but only in the page's white spaces, margins a crude dam breached when a book's watershed expands. Happening before your eyes. The river gods throw dice to decide on the first reader to float against his or her will. A throw of the dice stops no flood. Only drowned initiates need apply for the next crossing.

In objecting to her doctor's insistences, my mother, knowing she was rapidly fading anyway, argued against his harping upon her drinking eight glasses of water a day. "I've never had a stomach for the taste of water." Certainly that fit my memory of her staring at the Atlantic from a dirt-streaked window in Atlantic City muttering a curse toward the Atlantic, who was her grandfather, the tiring expanse of his impossible-to-navigate tyranny. This would be purely nostalgic wreckage—save for the fact she swims now in death's currents as she did in utero, with the discomfort of one for whom water can't dissolve an aftertaste of salty ash extracted from water-bloated husbands. Disappointing water, assuring no feminine purity, no Jewish baptism after the Fall.

Graffiti on the wall of the municipal water treatment plant: several high school sweetheart pacts sealed within hearts, a swastika. What filtration for those hormonal, political upsurges? A building my city pretends doesn't exist, as if water shouldn't wear such a public face for its treatment, left for the adolescents to deface, reconfigure, those swimming against the illusion of the bottomless well.

Claims adjuster stares at the spreading water stain covering the bedroom ceiling. It resembles Africa prior to its detachment from the larger land mass of a colonial mind-set. Scribbles in his notebook, aware of controlling the rhythm of a domestic drama, then reports, "Can't imagine the source or path. The water has a mind of its own." If he really heard his words' resonance, his insurance game would stop cold, he'd be out in my yard, dowsing stick in hand, for the hell of it.

Mushroom water, soot black & with a stench of ripe peat, result of cooking wild mushrooms of questionable identity, toxicity. Where the water was tossed on the lawn burned a saffron circle. The dog favored nibbling on grass within that circle as if lapping a dense gravy. When the dog died there was no question about where to bury it in the yard, a mere coincidence that that spot was where sprinklers couldn't reach, yet mushrooms sprout. Edible, they reduce in the pan to the Black Sea the Weinsteins supposedly left for good.

Holding in a rigidly upraised arm a glass of water before the nervous audience in a packed auditorium at an "alternatives" conference, anticipation palpable in the crowd, the speaker opened with "This will kill you faster than anything on earth!" The front rows looked anxious to line up to take the first draught to prove the point.

(Overheard uptown): "Bottled waters are virtually identical. Marketing demands novelty, so they have me design this bottle that looks like a million. It's the bottle, the label, we're pitching—not the water." Translates into: No way it's simply an illusion we're selling. It's the illusion that makes seeing through illusion a thrill. There's profit, loss, an unaccountable sense of loss when a bottled-water truck on Sunset runs out of gas while idling at a stuck red light, shamelessly advertising pathos of its come-hither mermaid icon on its dusty cab, aching for Neptune's tsunami thrust.

Part of a college's folklore: the famous rock star who visited the campus & came away impressed by a nineteenth-century water pump out of commission because vandals stole the handle. Enshrines the image in a radio hit. Out of which pour sweet waters of commercial success, an affrontery to the "liberal" administration & town antiquarians who treasure the broken pump as quaint Americana à la Rockwell, as in George Lincoln. Later discovered the pump handle in a corner of the President's office, someone with a thirst for keeping the revered broken broke.

In solidarity with the victims of The Camps, my aunt said a prayer before entering a shower, a curse upon departing.

Bless bathtubs for naked philosophers. Bless nakedness for demanding water dance in pursuit of a truth. Bless streaming words in pursuit of philosophies good to the last. Drop syllables until they splatter in microtonal pools. Bless watery guitar sounds from pressed-off-center Dub 45s, bass rumblings frogs would croak for.

Which rivers did Cézanne collect water from, & what were his methods of collecting, in order to arrive at his pearlized luminescent grays? Not a fruitless question. Any painter tapping such waters now would forestall the earth's death by a decade.

Billowing yellow sulfurous clouds filled the shower stalls, a fact of living where New York water ran stinking from every faucet. The sole redeeming bit of humor making the inconvenience tolerable was the thought that the water was also riddled with mercury from a nearby transformer plant. Daily living in the midst, mist of such currents ensured alchemical change, but most complained of undiagnosable headaches, the sense of a jawbone coming unhinged during heated debate, a maddening hypersensitivity to glare reflected in swimming pools and stagnant ponds.

"Because water erodes rock, the notion of a lasting Promised Land is meaningless as oceans rise." Although once considered a holy teacher, his disciples attributed his remark to senility. His wife shuddered at the extent of his mental decline, marking it by the time she had to throw a glass of cold water in his face to stop his ravings about being Jonah. His refusal to eat or drink until the day of the Messiah meant that she did his drinking for him, becoming for him his perfect image of the Dead Sea.

In the event of an emergency: one gallon of water per person per day, until the victims of such an emergency decide they wish to suicide by consuming sufficient water to do so, or wish to stop consuming water. In the event of such a death during an emergency, the ritual washing of the corpse can be performed using a variety of pleasant-tasting bottled or canned fruit juices. Should neither sufficient water or juice be available, saliva may be applied to the lips of the deceased. Those who can't bring themselves to apply spit upon the dead can be excused. This is only a test.

If American English could develop a vocabulary and way of thinking like Hopi, there would be one word meaning: "A stone is skimming across the water's surface, it is approaching dematerialization, there is no longer stone or water, the mind where the stone & water skim dissolves in sunlight, there is only the skimming movement when you breathe." Such a word remains timelessly current.

# Contributors

***Will Alexander*** is a Los Angeles poet whose books include *Towards the Primeval Lightning Field* and *Asia and Haiti*. His work has been published in the journals *Hambone, Callalloo,* and *Five Fingers Review,* among others, and his novella, *Alien Weaving,* is forthcoming.

***Antler,*** from Milwaukee, is the author of *Factory* and *Last Words.* His work also appears in many anthologies, including *The Soul Unearthed: Celebrating Wildness and Personal Renewal Through Nature.*

***Bob Braine*** is an installation artist and photographer whose projects investigate wildlife and ecosystems in urban environments, most recently those of New York City's East River and the Elbe River in Hamburg, Germany.

***Laynie Browne***'s recent books include *Rebecca Letters, The Agency of Wind,* and *Clepsydra.* Forthcoming is *Gravity's Mirror.* She lives in Seattle, where she teaches poetry in public schools and creative writing at the University of Washington at Bothell. She is also one of the curators of the Subtext Reading Series.

***Joseph Bruchac*** is a poet and storyteller whose work often reflects his Abenaki Indian ancestry. His writing has appeared in numerous publications, among them *American Poetry Review* and *Parabola,* and he has edited a number of highly praised literary anthologies. His latest book of poetry is *No Borders.*

***Malcolm de Chazal*** (1902–1981) was a writer, painter, and mystic from Mauritius, who earlier studied engineering in Louisiana. His *Sens-Plastique* was praised by Breton and Auden when it was published in Paris in 1948.

***Adam David Clayman*** is a documentary photographer residing in New York City. An ongoing major project of his involves the documenting of sacred sites in India.

***Anne Collet,*** one of the foremost experts on cetacean behavior, directs the Center for Research on Marine Mammals in France and the France-Europe Department of the National Center for the Study of Marine Mammals. In addition to many scientific publications, she is the author of a book of essays on her encounters with marine mammals, *Swimming with Giants*.

***Monique Crépault***'s ephemeral site-specific installations are sensitive to local ecologies and evoke eternal themes of humanity's relations to nature, as do her clay, cement, and plaster sculptures. She is based in Montreal, Canada.

***Hugh Dunkerley***'s poetry has been published widely in magazines and anthologies in the United Kingdom. Winner of an Eric Gregory award from the Society of Authors, his collection *Walking to the Fire Tower* was published by Redbeck Press.

***Matthew Ebinger*** is pursuing a bachelor of science degree in science, technology, and society with a concentration in environmental studies at the New Jersey Institute of Technology. He lives in northern New Jersey. The editors gratefully acknowledge his assistance in bringing this book into being.

***John Einarson*** is the editor of *Kyoto Journal*.

***Helen M. Ellis*** began photographing while working at a homeless drop-in center in New York City, which inspired one of her major projects and a continuing emphasis on portraiture in her work. Subsequent projects have taken her to Nicaragua, Ecuador, and Japan. Her work has been exhibited and published widely. She lives in New York City.

***Sally Gall*** lives in New York City, where she taught at the School of Visual Art and the International Center of Photography. Her photographs can be found in collections at the San Francisco Museum of Modern Art, the Museum of Fine Art at Houston, and the Bibliothèque Nationale of Paris. She is the author of *Water's Edge*.

***Sean Gillihan*** writes from Klamath Falls, Oregon. His work has appeared in a number of publications, including *Northern Lights, High Country News, Coe Review,* and *Rio Grande Review.* He has been awarded a Walden Residency Fellowship and an Oregon Literary Arts Fellowship.

***Jody Gladding***'s book *Stone Crop* appeared in the 1993 Yale Younger Poets Series. Her poetry has also appeared in *Grand Street, Paris Review,* and *Wild Earth.* She lives in Vermont, where she translates French for a living.

***Robert Grudin*** is the author of *Mighty Opposites, Book: A Novel, Time and the Art of Living, The Grace of Great Things,* and *On Dialogue.*

***Wilson Harris,*** one of the outstanding literary innovators of the twentieth century, is a writer of fiction, poetry, drama, and essays. He was born in Guyana in 1921 and has lived in London for the past forty years. His novels date from *Palace of the Peacock* (1960) to *Jonestown* (1996), and a collection of his essays was recently published.

***Freeman House,*** a former commercial salmon fisherman, is cofounder of the Mattole Watershed Salmon Support Group and the Mattole Restoration Council. He is the author of *Totem Salmon: Life Lessons from Another Species.*

***George Keithley*** is the author of seven books of poetry, two plays, and numerous short stories and essays. His award-winning epic, *The Donner Party,* was a Book-of-the-Month Club selection and has been adapted as a play and an opera.

***Tom LeClair*** is the author of the novel *Passing Off* and a collection of ecocriticism, *The Art of Excess.* His work has also appeared in *TriQuarterly, Paris Review, Witness,* and *Fiction International.*

***Margaret McCarthy*** works as a photographer in New York City. Her landscape photographs and pieces based on Celtic mythology have been exhibited widely.

***Arno Rafael Minkkinen*** is a professsor of art at the University of Massachusetts Lowell. Water has been a common element in his self-portrait photographs for nearly three decades now. Minkkinen was awarded the first *Scritture d'Acqua* Prize in Italy in 1996. His recent books are *Waterline* and *Body Land.*

***David Morse*** is an activist-writer. His most recent book is a novel, *The Iron Bridge.* His web site is www.david-morse.com.

***Melissa Nelson*** is the director of the Cultural Conservancy, a San Francisco organization dedicated to forging links between environmentalism and Native American communities.

***John P. O'Grady*** is the author of *Grave Goods* and *Pilgrims to the Wild.* He teaches at Allegheny College in Meadville, Pennsylvania.

***Kristin Ordahl*** started making video work based on her paintings in 1994. Her work has been shown at the Holly Solomon Gallery and David Zwirner Gallery in New York City. She lives and works in Brooklyn, New York.

***Bivash Pandav*** has dedicated himself to the research and conservation of olive ridleys of Orissa, India, for the past five years. His untiring efforts have helped focus attention on the mass mortality of these sea turtles due to illegal trawl fishing and has inspired many conservation initiatives. He works at the Wildlife Institute of India in Dehradun.

***Raimundo Panikkar,*** a scholar of science, philosophy, and theology, is the author of *The Silence of God: The Answer of the Buddha.* He lives in Spain.

***Ricardo Pau-Llosa*** has published four books of poetry: *Sorting Metaphors, Bread of the Imagined, Cuba,* and *Vereda Tropical.* He has also published widely on Latin American art.

***Octavio Paz*** (1914–1998), the late Mexican poet and critic, was awarded the 1990 Nobel Prize for literature. His poetry includes *The Violent Season* (1958) and *Configurations* (1971). Among his prose works are *The Labyrinth of Solitude* (1950), *Children of the Mire: Modern Poetry from Romanticism to the Avant-Garde* (1974), and *The Monkey Grammarian* (1981).

***Jaanika Peerna*** is an artist and educator from Estonia, now living in Cold Spring, New York.

***Sidney Perkowitz*** is the Charles Howard Candler Professor of Physics at Emory University in Atlanta, Georgia. He is the author of *Empire of Light* and *Universal Foam,* about the science of foam—from cappuccino to galaxies in space.

***D. L. Pughe*** is a writer and artist living in Berkeley, California. Her recent essays have appeared in *When Pain Strikes* and *Searchlight: Consciousness at the*

*Millenium,* and passages from her work *A Philosophy of Clean* can be found in *Nest* magazine and *The New Earth Reader: The Best of Terra Nova.*

***C. L. Rawlins*** lives a wind-blown life in Wyoming. A field hydrologist, he won the U.S. Forest Service National Primitive Skills Award for scientific fieldwork in the Wind River Range. He has written two nonfiction books, *Sky's Witness: A Year in the Wind River Range* and *Broken Country: Mountains and Memory,* and two books of poetry: *A Ceremony on Bare Ground* and *In Gravity National Park.* "A Myth for Sofie" is drawn from a book in progress, titled *Robyn and Sleep.*

***Pamela Ryder*** is a nurse practitioner in a New York City hospital. Her work has been published in *Quarterly, Prairie Schooner, Black Warrior Review,* and *American Writing.*

***Eva Salzman*** is a Royal Literary Fund Fellow at Ruskin College, Oxford. She is the author of two books of poetry, *Bargain with the Watchman* and *The English Earthquake,* and recently completed two opera librettos for the English National Opera studio.

***Kartik Shanker*** thinks he is becoming an ecologist, having worked in remote forests on rodent communities and walked windy beaches in search of sea turtles. At heart, though, he is probably just a writer. He lives in India.

***Ted Steinberg*** is associate professor of history and law in the Department of History at Case Western Reserve University. He is the author of *Acts of God: The Unnatural History of "Natural" Disasters in America.*

***Dan Stryk*** is a professor of world literature and creative writing at Virginia Intermont College. He is the author of five collections of poetry, and his poems and prose parables have appeared in numerous literary and cultural journals, such as *Southern Humanities Review, TriQuarterly,* and *Tricycle: The Buddhist Review.*

***Linda Tatelbaum*** homesteads in midcoast Maine and teaches at Colby College. She is the author of *Carrying Water as a Way of Life: A Homesteader's History* and *Writer on the Rocks: Moving the Impossible.*

***Jerry Uelsmann*** has been graduate research professor of art at University of Florida since 1974. His work can be viewed in the Metropolitan Museum in New York and the Royal Photographic Society in London, as well as in other major collections. The most recently published collections of his photographs are *Museum Studies* and *Approaching the Shadow.*

***Marilyn Ulvaeus*** is a social and environmental activist living near Santa Barbara, California. She loves to travel. And if you're looking for her, you'll probably find her in or around a body of water.

***Arthur Versluis*** teaches at Michigan State University where he is also editor-in-chief of *Esoterica,* an electronic journal at http://www.esoteric.msu.edu/, devoted to the academic study of Western esotericism. His books include *Wisdom's Children: A Christian Esoteric Tradition* and *Island Farm,* an account of his experiences working on his family's commercial farm in Michigan.

***Peter Warshall***'s work centers on conservation and conservation-based development. He runs a consulting firm, was an adjunct research scientist with the Office of Arid Lands Studies (University of Arizona), lectures, writes, and is the editor of the *Whole Earth Review.*

***Eliot Weinberger*** is a writer, poet, and translator. He is the author of the essay collections *Works on Paper* and *Outside Stories,* and he has translated works of Octavio Paz, Jorge Luis Borges, and Bei Dao. He lives in New York City.

***Norman Weinstein*** is a poet and critic whose books include *A Night in Tunisia: Imaginings of Africa in Jazz* and *Suite: Orchid Ska Blues.* He writes about music and technology for a variety of publications. His script "In Tune with Nature," based on a *Terra Nova* issue, was produced by WGBH Radio Boston for its Sound and Spirit series.

***Irving Weiss***'s most recent books are collections of visual poems, *Number Poems* and *Visual Voices: The Poem As a Print Object*; and *Reflections on Childhood,* with Anne D. Weiss. His visual and word poems, essays, and translations have been published in many anthologies and magazines.

***Marie Wilkinson*** and ***Cyril Christo*** work together as photographers collecting images and stories of this changing world. Marie is an architect, and Cyril is a documentary filmmaker and poet. They live in Santa Fe, New Mexico.

***Christopher Woods*** is a native Texan writer of fiction, nonfiction, poetry, and plays. His work has appeared in numerous literary journals, including *Columbia, Southern Review,* and *New England Review.* He is the author of the novel

*The Dream Patch* and teaches writing in Houston at the Women's Institute and Rice University.

***Gayle Wurst*** is guest professor at the University of Bordeaux, France, where she teaches translation and American literature. The author of *Voice and Vision,* a work on Sylvia Plath, and the translator of Anne Collet's *Danse avec les baleines,* her numerous publications on contemporary writers have appeared in French and English.

***David Rothenberg*** is a philosopher and musician. He is associate professor of philosophy at the New Jersey Institute of Technology and the founder of *Terra Nova,* the annual book series on the culture of nature. His latest CD is *Before the War,* and his forthcoming book, *Sudden Music: Improvisation, Art, Nature,* will be published in the fall of 2001.

***Marta Ulvaeus*** is the associate director of continuing Terra Nova projects at the New Jersey Institute of Technology. She was previously an editor of *TDR: The Drama Review,* at New York University, where she did graduate work in performance studies. For years she could be heard spinning discs on community radio stations KPFA and KDVS in California.

***Christopher Funkhouser,*** editor of We Press and *Newark Review* and poetry editor for *Terra Nova,* has produced publications by Kamau Brathwaite and Amiri Baraka. He lives on Staten Island, New York, and teaches at the New Jersey Institute of Technology.

# Sources

Bob Braine, "The Bronx River," from *Two Waters.* Reprinted with permission from the author and courtesy of Galerie für Landschaftskunst, Hamburg.

Malcolm de Chazal, excerpts from *Plastic Sense,* translation of *Sens-plastique,* by Irving Weiss (New York: Sun Press, 1979). Reprinted courtesy of Irving Weiss.

Anne Collet, "Swimming with Children and Three Hundred Dolphins," translated by Gayle Wurst, from *Swimming with Giants: My Encounters with Whales, Dolphins, and Seals* (Milkweed, 2000), originally published as *Dance avec les baleines* (Plon, 1998). Text © by Anne Collet and translation © 2000 by Gayle Wurst. Reprinted with permission from Milkweed Editions.

Wilson Harris, "The Music of Living Landscapes," from *Selected Essays of Wilson Harris: The Unfinished Genesis of the Imagination* (London: Routlege, 1999), Andrew Bundy, ed. Reprinted by permission of Wilson Harris.

Freeman House, "In Salmon's Water" from *Totem Salmon* © 1999 by Freeman House. Reprinted by permission of Beacon Press, Boston.

Raimundo Panikkar, "The Hymn of the Origins" from *The Vedic Experience* (University of California Press, 1977), reprinted by permission of Raimundo Panikkar.

Octavio Paz, "My Life with the Wave" from *Eagle or Sun?* translation © 1976 by Octavio Paz and Eliot Weinberger. Reprinted by permission of New Directions Publishing Corp. and Lawrence Pollinger Limited.

Sidney Perkowitz, "The Rarest Element" from *The Sciences,* (January–February 1999). Reprinted by permission of *The Sciences,* 2 East 63rd Street, New York, NY 10021.

Eva Salzman, "There's a Lot of Room," from *New Writing 9* (Vintage/British Council, 2000), A. L. Kennedy and John Fowles, eds. Reprinted by permission of the author.

Ted Steinberg, "Morton Salt Disaster" adapted from *Acts of God: The Unnatural History of "Natural" Disasters in America* (New York: Oxford University Press) © 2000 Ted Steinberg. Reprinted by permission of the publisher.

Linda Tatelbaum, "The Lessons of the Well" from *Writer on the Rocks: Moving the Impossible* (About Time Press, 2000). Reprinted by permission of the author.